FLORA OF TROPICAL EAST AFRICA

THYMELAEACEAE

B. Peterson

(University of Göteborg)

Trees or shrubs (sometimes lianes), rarely perennial herbs, very rarely annuals. Stems and branches with tough cortical, often shining, fibres. Leaves alternate or opposite, sessile or shortly petiolate, simple, entire, small needle-like, 1-nerved, to large flat, pinnately nerved, without stipules, herbaceous or coriaceous, sometimes glandular-punctate. Inflorescences terminal or axillary, sessile or pedunculate, racemose (spikes, fascicles, umbels or heads), flowers rarely solitary, often with deciduous or persistent bracts. Flowers bisexual, polygamous or dioecious, regular or rarely slightly irregular, sweetly scented at night. Calyx (hypanthium, perianth or receptacle) tubular or funnel-shaped, sometimes articulated above the ovary, usually coloured, often petal-like; lobes (3–)4–5(–6), usually imbricate, equal or often with the two interior slightly smaller. Petals (petaloid appendages or scales) generally inserted in the throat of the calyx-tube, equal or double the number of the calyx-lobes, well developed, entire or divided, often reduced to small fleshy glands or lacking. Stamens as many or twice as many as the calyx-lobes (rarely reduced to 2 or 1), in 1–2 whorls, the upper whorl antisepalous; anthers with short filaments or sessile, 2-thecous, usually introrse, rarely extrorse or horseshoe-like, opening by slits lengthwise; pollen globose, usually polyforate with " croton-pattern ". Ovary superior, 1–2(rarely 4–12)-locular, sessile or shortly stipitate; ovule 1 in each locule, usually pendulous, anatropous; style filiform, sometimes very short, terminal or lateral; stigma usually capitate, sometimes papillate. Disc hypogynous, membranous or fleshy, annular, cupular or scale-like, sometimes minute or lacking. Fruit often a berry, sometimes a nut, drupe or loculicidal capsule, usually enclosed in the base of the persistent calyx-tube. Seed usually with a caruncle-like appendage, outer coat thin or crustaceous, usually black, with or without endosperm; embryo straight; cotyledons flat or thickened, narrow or broad. Most thymelaeaceous plants contain toxic substances.

A family of some 50 genera and 600 species widespread in tropical and subtropical regions, also temperate.

With small modifications the system published by W. Domke in Bibliotheca Botanica 27(111):1–151 (1934) is followed here. All genera represented in East Africa belong to the subfamily *Thymelaeoideae*.

Concerning the interpretation of the complex structure of the calyx-tube and the petals see J. Léandri in Ann. Sci. Nat., Bot. sér. 10, 12:125–237 (1930) and K. Heinig in Amer. Journ. Bot. 38:113–132 (1951).

Regarding the distribution of *Thymelaeaceae* within tropical Africa see G. Aymonin in C.R. Soc. Biogéogr. Paris 365:6–21 (1965).

Stamens twice as many as the calyx-lobes :
  Calyx-tube not articulated above the ovary :
    Petals present :
      Flowers solitary, in terminal cymes or axillary :
        Petals well developed, much longer to shorter
          than the calyx-lobes . . . . 1. **Dicranolepis**

Petals small, glandular:
Petals forming an entire or lobed, often
ciliate ring . . . . . 2. **Synaptolepis**
Petals in the shape of 10 minute glands . 3. **Craterosiphon**
Flowers in terminal heads . . . . 6. **Gnidia***
Petals absent:
Flowers axillary; calyx-tube funnel-shaped . 3. **Craterosiphon**
Flowers terminal; calyx-tube cylindric:
Flowers in racemes or umbels . . . 4. **Peddiea**
Flowers in heads . . . . . 5. **Dais**
Calyx-tube articulated above the ovary . . 6. **Gnidia**
Stamens as many as the calyx-lobes . . . 7. **Struthiola**

1. DICRANOLEPIS

Planch. in Hook., Ic. Pl. 8, t. 798 (1848); H.H.W. Pearson in F.T.A. 6(1): 238
(1910); Domke in Bibl. Bot. 27(111): 121 (1934); Aymonin in Fl. Gabon 11:
64 (1966) & in Fl. Cameroun 5: 9 (1966); A. Robyns in F.A.C. Thymelaeaceae:
8 (1975)

Shrubs or small trees. Branches spreading horizontally, young twigs
glabrous to pubescent. Leaves alternate, petiolate; blade oblong or elliptic,
usually unequal-sided, long-acuminate, basally cuneate to obtuse, chartaceous, glabrous or pubescent, pinnately nerved. Inflorescence axillary.
Flowers 5-merous, subsessile or shortly pedicelled, solitary or 2–4 or clustered
in the leaf axils, bracteate. Calyx-tube cylindric, usually elongate, ± pubescent, sometimes glandular; lobes 5, usually pubescent outside. Petals 5,
shorter or longer than the calyx-lobes, often divided nearly to the base, entire,
laciniated or irregularly shaped at the top, white or very pale yellow. Stamens 10, in two whorls in the throat of the calyx-tube; episepalous filaments
slightly longer than the alternisepalous; anthers included or exserted. Ovary
shortly stipitate, 1-locular, glabrous or pubescent; disc cup-shaped, ± deeply
lobed, fleshy or membranous; style slender; stigma clavate or capitate, included or exserted. Fruit drupe-like, spherical or elongate, brown, red or
orange, enclosed in the persistent base of the calyx-tube. Seed with thin
testa often fixed to the hard pericarp.

Tropical African genus of about 15 species.

Calyx-tube 20–30 mm. long:
Petal-lobes narrowly oblong, irregularly divided by
deep incisions . . . . . . 1. *D. incisa*
Petal-lobes spathulate, almost entire or irregularly
denticulate at the top . . . . 2. *D. usambarica*
Calyx-tube less than 15 mm. long . . . . 3. *D. buchholzii*

1. **D. incisa** *A. Robyns* in B.J.B.B. 45: 97, fig. 1 (1975) & in F.A.C. Thymelaeaceae: 17, t. 3 (1975). Type: Zaire, Walikale, Mifuti, *Léonard* 1856 (BR,
holo.)

Shrub or small tree up to 3 m. high. Young branches densely pubescent,
later glabrous. Petiole 1–3 mm. long, glabrous or pubescent; leaf-blade
oblong, unequal-sided, 40–80(–93) mm. long, 20–35 mm. wide, acuminate
apex 8–12(–15) mm. long, chartaceous, glabrous above, sparsely adpressed
pubescent beneath (especially young leaves); midrib and lateral nerves prominent beneath. Flowers silvery white, usually solitary; pedicel 1–3 mm.
long, pubescent like the small triangular bracts. Calyx-tube 20–25 mm. long,

* *Gnidia glauca* p. 32

FIG. 1. *DICRANOLEPIS USAMBARICA*—**1**, flowering branch, × ⅔; **2**, flower, × 2; **3**, same, showing variation in indumentum, × 2; **4**, longitudinal section of flower, × 2; **5**, pistil and disc, × 3; **6**, details of same, × 10; **7**, longitudinal section of ovary and disc, × 10; **8**, part of fruiting branchlet, × 2. 1, 2, 4–7, from *Harris* 618; 3, 8, from *Drummond & Hemsley* 1832.

1–2 mm. wide, densely white pubescent; lobes ovate-lanceolate, 10–13 mm. long, 2·5–3·5 mm. wide, finely pubescent on both sides. Petals white, at least slightly exceeding the calyx-lobes, divided to the base, 10–14 mm. long, 2–3 mm. wide, oblong, at the upper part irregularly divided by deep incisions. Stamens exserted, filaments 5–8 mm. long, anthers 1–2 mm. long. Ovary shortly stipitate, glabrous; disc 1–2 mm. high, slightly and irregularly lobed; style 25–35 mm., somewhat exserted; stigma papillate.

UGANDA. Bunyoro District: Budongo Forest, Sept. 1933 (fl.), *Eggeling* 1373! & Sept. 1940 (fl.), *Eggeling* 4050!; Masaka District: Tero Forest, (fl.), *Fyffe* 154!
DISTR. U2, 4; Zaïre
HAB. Rain-forest; 1200–1300 m.

2. **D. usambarica** *Gilg* in E.J. 19 : 272 (1894) & in P.O.A. C : 284, t. 32/A–C (1895); H.H.W. Pearson in F.T.A. 6(1) : 241 (1910); T.T.C.L. : 607 (1949). Type: Tanzania, Usambara Mts., Mbaramu, *Holst* 2489 (B, holo. †, HBG, K, iso.!)

Slender shrub 1–4·5 m. or small tree up to 6(–15) m. high. Young branches pilose or pubescent, later glabrous. Petiole 2–4 mm. long, glabrous to pubescent; leaf-blade oblong or elliptic-oblong, ± irregular, 50–85(–110) mm. long, 20–38(–58) mm. wide, acuminate apex 5–10 mm. long, chartaceous, glabrous above, sparsely hairy beneath, midrib and lateral nerves prominent on both sides, light green beneath. Flowers white or yellowish white, solitary or paired; pedicel 1–2 mm. long, slightly pubescent like the small triangular bracts. Calyx-tube 20–30 mm. long, 1–2 mm. wide, almost glabrous to densely puberulous or tomentose; lobes ovate to ovate-oblong, 6–12 mm. long, 2–4·5 mm. wide, sericeous on the outside or pubescent on the margins, puberulous or almost glabrous on the inside. Petals equal to or slightly longer than the calyx-lobes, divided almost to the base; lobes spathulate, 6–12(–14) mm. long, 1·5–4 mm. wide, almost entire to irregularly denticulate at the top. Filaments 0·5–2 mm. long; anthers 1–1·5 mm. long, exserted. Ovary shortly stipitate, glabrous; disc 1–2 mm. high, deeply lobed; style exserted, stigma papillate. Fruit red, 10–12 mm. long, 7–9 mm. wide. Fig. 1, p. 3.

KENYA. Teita District: Teita Hills, Mbobolo Hill, Sept.–Oct. 1938 (fl., fr.), *Joanna* in *C.M.* 9000! & July 1969 (fl.), *Faden, Evans & Wolf* 69/830! & Mraru Mt., 9 Feb. 1966 (fl., fr.), *Gillett, Burtt & Osborn* 17154!
TANZANIA. Lushoto District: Shagayu Forest Reserve, Goka, 30 Sept. 1964 (fr.), *Mgaza* 607!; Morogoro District: Nguru Mts., Manyangu Forest, Liwale valley, 27 Mar. 1953 (fl., fr.), *Drummond & Hemsley* 1832! & Uluguru Mts., Mwere valley, 26 Sept. 1970 (fl.), *Harris et al.* 5096!
DISTR. K7; T3, 6; not known elsewhere
HAB. Rain-forest; 600–1800 m.

SYN. *D. usambarica* Gilg var. *nana* Domke in N.B.G.B. 11 :671 (1932); T.T.C.L. :607 (1949). Type: Tanzania, Ulanga District, Mahenge, Sali, *Schlieben* 2205 (B, holo. †, B, G, HBG, S, iso. !)

NOTE. The indumentum of the calyx-tube in this species varies from glabrous to densely puberulous or tomentose. There is no geographical distinction between the extremes though the glabrous form seems to be more common from the Uluguru Mts. and southwards.

3. **D. buchholzii** *Engl. & Gilg* in E.J. 19 : 273 (1894); H.H.W. Pearson in F.T.A. 6(1) : 241 (1910); Domke in Bibl. Bot. 27(111), t. 1/7 (1934); F.W.T.A., ed. 2, 1 : 173 (1954); Aymonin in Fl. Gabon 11 : 78, t. 11/1–3 (1966) & in Fl. Cameroun 5 : 30, t. 4/1–3 (1966); Archangelsky in Bot. Zhurn. 51, t. 4/3 (1966) & in Kuprianova, Pollen Morphology: 134, t. 3/10 (1971); A. Robyns in F.A.C. Thymelaeaceae: 9 (1975). Type: Cameroun, Mungo, *Buchholz* (B, holo. †)

Shrub or rarely small tree up to 4(–9) m. high. Branches pubescent, later glabrous. Petiole 1–4 mm. long, glabrous or pubescent; leaf-blade oblong or ovate-oblong, unequal sided, 40–85(–100) mm. long, 20–30(–45) mm. wide, acuminate apex 10–16(–20) mm. long, chartaceous, glabrous on both sides or slightly hairy beneath, often pubescent along the midrib, venation prominent beneath, dark green above, brownish beneath. Flowers white, solitary or 2–4(–10); pedicel none or 1–2 mm. long, slightly pubescent. Calyx-tube (6–)8–10(–15) mm. long, ± 1 mm. wide, glabrous or slightly pubescent; lobes narrowly ovate, inner 3–4 mm. long, outer 4–5 mm. long, ± 1 mm. wide, glabrous or pubescent outside, glabrous or puberulent inside. Petals usually much shorter than the calyx-lobes, divided nearly to the base, the 10 lobes lanceolate or linear, 1·5–3·5 mm. long, 0·5–1·2 mm. wide. Anthers ± 1 mm., slightly exserted. Ovary glabrous; disc 1 mm. high; style included or slightly exserted. Fruit 7–13 mm. long, 5–10 mm. wide, glabrous or slightly pubescent.

UGANDA. Kigezi District: Ishasha Gorge, Feb. 1949 (fl.), *Purseglove* 2710! & 5 Aug. 1971 (fl.), *Katende* 1259! & Kayonza Forest Reserve, 5 Aug. 1960 (fl.), *Paulo* 658!
DISTR. **U**2; Cameroun, Gabon, Zaïre, Central African Empire
HAB. Evergreen forest; 1400–1800 m.

SYN. *D. disticha* Planch. var. *parviflora* Engl. in E.J. 7: 337 (1886). Type: Cameroun, Mungo, *Buchholz* (B, holo.†)
    D. oligantha Gilg in E.J. 19: 274 (1894); Staner in B.J.B.B. 13: 328 (1935); F.P.N.A. 1: 652 (1948). Type: Gabon, Munda, Sibange Farm, *Soyaux* 22 (B, holo.†, K, P, iso.!)
    D. thonneri De Wild. & Th. Dur. in B.S.B.B. 38(2): 114 (1899) & Pl. Thonn. Cong.: 29, t. 10 (1900); H.H.W. Pearson in F.T.A. 6(1): 245 (1910). Type: Zaire, Boyangi, near Ndobo, *Thonner* 62 (BR, holo.!, K, iso.!)
    D. parviflora H.H.W. Pearson in F.T.A. 6(1): 244 (1910) & in K.B. 1910: 340 (1910). Type: Zaire, Boyangi, near Ndobo, *Thonner* 62 (K, holo.!, BR, iso.!)
    D. mildbraedii Gilg in Z.A.E.: 577 (1913). Type: Zaire, near Muera, *Mildbraed* 2171 (B, holo.†, HBG, iso.!)
    D. batesii S. Moore in J.B. 57: 117 (1919). Type: Cameroun, Bitye, *Bates* 692 (BM, holo.!)
    ?*D. humillima* Engl. in V.E. 3(2): 638 (1921), *nom. nud.*

2. SYNAPTOLEPIS

Oliv. in Hook., Ic. Pl. 11: 59, t. 1074 (1870); H.H.W. Pearson in F.T.A. 6(1): 245 (1910); Domke in Bibl. Bot. 27(111): 122 (1934)

Small shrubs or woody climbers. Branches slender often twining; bark lenticellate. Leaves alternate or opposite, petiolate; leaf-blade ovate to elliptic-lanceolate, membranous to subcoriaceous, glabrous, lateral nerves parallel, spreading, prominent beneath, margin thickened. Inflorescence terminal cymes or axillary. Flowers 5-merous, solitary or fascicled, bracteate; pedicel sometimes glandular. Calyx-tube cylindric, glabrous. Petals forming an erect, entire or lobed, often ciliate, ring at the base of the calyx-lobes. Stamens 10, in two whorls in the upper part of the calyx-tube, the episepal usually slightly exserted. Ovary shortly stipitate, 1-locular; surrounded at the base by a minute, ± deeply lobed disc; style slender, stigma capitate, papillate. Fruit dry, ovoid, enclosed in the persistent base of the calyx-tube. Seeds without endosperm.

Four or five species in tropical Africa and one from Madagascar.

Inflorescence terminal; petals with stiff hairs .   . 1. *S. alternifolia*
Inflorescence axillary; petals without stiff hairs   . 2. *S. kirkii*

1. **S. alternifolia** *Oliv.* in Hook., Ic. Pl. 12: 8, t. 1194 (1876); H.H.W. Pearson in F.T.A. 6(1): 246 (1910); T.T.C.L.: 612 (1949); Peterson in Bol. Soc. Brot., sér. 2, 33: 218 (1959). Type: Tanzania, Kilwa, *Kirk* (K, holo.!)

Shrub or woody climber up to 10 m. high. Branches slender, glabrous; bark reddish brown or black, lenticellate. Leaves alternate or opposite; petiole 1·5–3 mm. long; leaf-blade elliptic or ovate, (12–)15–45(–63) mm. long, (6–)10–25(–32) mm. wide, acuminate to truncate at the apex, cuneate or shortly narrowed at the base, sometimes slightly undulate, membranous to subcoriaceous, glabrous. Flowers white, pale yellow or cream, in terminal pedunculate cymes; bracts 1–2 mm. long, ciliate, falling off rather early. Calyx-tube 12–18 mm. long, glabrous; lobes ovate, 2–3·5 mm. long, 1–2 mm. wide, glabrous. Petals with stiff hairs, ± 0·5 mm. long. Stamens of the upper whorl slightly longer than the petals; anthers 1 mm. long. Ovary with a few hairs at the top or glabrous, surrounded by a whorl of hairs at the base; style 8–14 mm. long. Fruit glabrous, yellow to reddish orange, 12–23 mm. long, 8–12 mm. wide. Fig. 2/1–6.

TANZANIA. Lushoto District: E. Usambara Mts., Kisiwani, 25 May 1943 (fr.), *Greenway* 6688!; Morogoro District: Magadu, 24 Oct. 1957 (fl.), *Welch* 410!; Tunduru District: 11 km. E. of Songea District boundary, 21 Dec. 1955 (fl.), *Milne-Redhead & Taylor* 7735!
DISTR. **T**3, 6–8; Malawi, Mozambique, Rhodesia
HAB. Riverine forest and thicket, *Brachystegia* woodland, secondary bushland and wooded grassland; 50–1100 m.

SYN. *S. longiflora* Gilg in E.J. 19: 276 (1894); H.H.W. Pearson in F.T.A. 6(1): 246 (1910); T.T.C.L.: 612 (1949); Peterson in Bol. Soc. Brot., sér. 2, 33: 218 (1959). Type: Mozambique, Zambesia, Macusi (Macuri), *Carvalho* (B, holo.†, COI, iso.!)

NOTE. There is at Kew a specimen by *Kirk* labelled " Zanzibar ? ". So far no other collections of this species have been made in Zanzibar I. and an imprecise labelling might be suspected.

2. **S. kirkii** *Oliv.* in Hook., Ic. Pl. 11: 59, t. 1074 (1870); H.H.W. Pearson in F.T.A. 6(1): 247 (1910); Domke in Bibl. Bot. 27(111), t. 2/12 (1934); T.S.K. ed. 2: 18 (1936); K.T.S.: 557 (1961). Type: Zanzibar, *Kirk* 37 (K, lecto.!)

Small shrub, most often climbing, up to 4 m. Branches divaricate, twining, with blackish bark and numerous bright lenticels, young shoots sometimes glandular. Leaves opposite, rarely alternate; petiole 1–2 mm. long; leaf-blade ovate to elliptic-lanceolate, (14–)24–38(–52) mm. long, (9–)14–20(–25) mm. wide, acute or apiculate, subcoriaceous, glabrous; lateral nerves parallel. Flowers white or cream, axillary, solitary or fascicled, pendulous; pedicel 2–10(–14) mm. long, usually glandular; bracts ± 1 mm. long, ciliate. Calyx-tube 11–16 mm. long, glabrous; lobes oblong, 4–5 mm. long, 1·5–2·5 mm. wide, inner slightly smaller, glabrous. Petals not ciliate. Stamens of the upper whorl slightly exceeding the petals. Ovary glabrous; style 5–8 mm. long. Fruit glabrous, orange, 11–14(–18) mm. long, 6–9 mm. wide. Fig. 2/7–12.

KENYA. Tana River District: Mambosasa, Feb. 1929 (fl.), *R.M. Graham* 1805!; Kilifi District: 5 km. beyond Marikebuni on Malindi–Marafa road, 16 Nov. 1961 (fl., fr.), *Polhill & Paulo* 776!; Kwale District: near Kwale–Tanga road, end of Mrima road, 8 Sept. 1957 (fl.), *Verdcourt* 1931!
TANZANIA. Tanga District: Gombero Forest Reserve, 14 Nov. 1964 (fl.), *Mgaza* 669!; Uzaramo District: 9.5 km. E. of R. Ruvu, 27 Nov. 1955 (fl.), *Milne-Redhead & Taylor* 7440!; Kilwa, Aug. 1873, *Kirk*!; Zanzibar I., Mazizini, 21 Nov. 1959 (fl.), *Faulkner* 2403! & Mbweni, 9 Jan. 1931 (fl., fr.), *Vaughan* 1068!; Pemba I., Mkoani road, 15 Aug. 1929 (fl.), *Vaughan* 506!
DISTR. **K**7; **T**3, 6, 8; **Z**; **P**; Mozambique, South Africa (Natal)

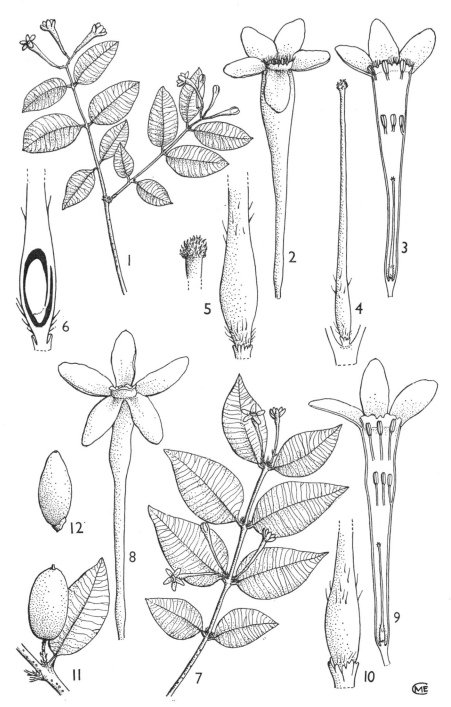

FIG. 2. *SYNAPTOLEPIS ALTERNIFOLIA*—**1,** flowering branch, × ⅔; **2,** flower, × 4; **3,** longitudinal section of flower, × 4; **4,** pistil and disc, × 10; **5,** details of same, × 20; **6,** longitudinal section of ovary, × 20. *S. KIRKII*—**7,** flowering branch, × ⅔; **8,** flower, × 4; **9,** longitudinal section of flower, × 4; **10,** ovary and disc, × 20; **11,** part of fruiting branchlet, × 1; **12,** seed, × 2. 1, from *Pawek* 5076; 2–6, from *Pawek* 6531; 7–10, from *Milne-Redhead & Taylor* 7440; 11, 12, from *Faulkner* 2114.

HAB. Lowland dry evergreen forest, *Brachystegia* woodland, coastal and secondary bushland or thicket; 0–450(–900) m.

SYN. ?*S. pachyphylla* Gilg in P.O.A. C: 284 (1895); H.H.W. Pearson in F.T.A. 6(1): 246 (1910); T.T.C.L.: 612 (1949). Type: Tanzania, Uzaramo District, Rukinga, *Stuhlmann* 6897 (B, holo.†)
      ?*S. macrocarpa* Gilg in P.O.A. C: 284 (1895); H.H.W. Pearson in F.T.A. 6(1): 247 (1910); T.T.C.L.: 612 (1949). Type: Tanzania, Uzaramo District, Vikindu, *Stuhlmann* 6098 (B, holo.†)
      ?*S. bussei* Gilg in V.E. 1(1): 390 (1910); T.T.C.L.: 612 (1949), *nom. nud.*, probably based on Tanzania, Kilwa District, Matumbi highlands, *Busse* 3107 (B†, EA !)
      [*S. longiflora* sensu K.T.S.: 558 (1961), quoad specim. cit., *non* Gilg]

NOTE. A specimen at Kew (Zanzibar?, *Kirk* !) named by Gilg as *S. pachyphylla* Gilg fits well within the variability of *S. kirkii*. In the original description of *S. pachyphylla* Gilg has stated " floribus terminalibus " but Kirk's collection has axillary flowers. The type specimen is lost and I have seen no isotypes but it appears to me that this species and *S. kirkii* are conspecific.
      S. macrocarpa was distinguished from *S. kirkii* only by the larger fruits (18 × 6–7 mm.) and several specimens at EA from Berlin so named are clearly *S. kirkii*. Although no modern fruits have been seen quite this size it is unlikely that more than one species is involved. The type locality was cleared of forest some 10 years ago and only partial regeneration has occurred.
      Spjut 3908 (NA !) from Kenya, Kilifi District, E. of Jilore Forest Station, has glands not only on the pedicels but also on the calyx-tubes and lobes.
      The fig. 81/F–J in E. & P. Pf. III. 6A: 231 (1894) does not represent *S. kirkii* but *S. oliverana* Gilg (" *oliveriana* ").

3. CRATEROSIPHON

Engl. & Gilg in E.J. 19: 275 (1894) & in E. & P. Pf. III. 6A: 233 (1894); Domke in Bibl. Bot. 27(111): 122 (1934); Aymonin in Fl. Gabon 11:54 (1966) & in Fl. Cameroun 5:35 (1966)

Lianes, climbing shrubs or small trees with twining branches; twigs with conspicuous lenticels, glabrous. Leaves opposite, subopposite or rarely alternate, petiolate; blade elliptical to ovate, acuminate, chartaceous to coriaceous; lateral nerves spreading almost at right angles. Inflorescence axillary. Flowers 5-merous, subsessile or pedicelled, solitary or 2–4, with small, sometimes ciliate, bracts. Calyx-tube funnel-shaped, glabrous outside, glabrous or hairy inside, somewhat fleshy, persistent; lobes 5, equal to the tube or much shorter. Petals absent or 10 minute glands inserted on the throat of the tube. Stamens 10 in two whorls, inserted in the throat of the tube. Ovary sessile or subsessile, 1-locular, glabrous or slightly pubescent or pilose at the top; disc cup-shaped or lobed, membranous; style terminal or sublateral; stigma capitate, finely papillose. Fruit a drupe with sclerified pericarp, included in the persistent calyx-tube. Grain oblong, fusiform.

About 9 species, all tropical African, mainly in the rain-forest regions of the centre and west.
A genus very closely related to *Synaptolepis*, distinguished however by the petals which are rudimentary or lacking. Most of the species are not yet adequately known. Further and better material, with flowers and fruits, is desired.

Calyx-tube glabrous inside; stamens in two whorls ± 0·5 mm. apart . . . . . . 1. *C. beniense*
Calyx-tube hairy inside; stamens in two whorls ± 3 mm. apart . . . . 2. *C. sp. A*

1. **C. beniense** *Domke* in N.B.G.B. 11: 352, fig. 8/1, 5, 6 (1932); Staner in B.J.B.B. 13: 334 (1935); Aymonin in Fl. Cameroun 5: 40 (1966); Robyns in F.A.C. Thymelaeaceae: 21 (1975). Type: Zaire, Beni, *Mildbraed* 2252 (B, holo. †, HBG, lecto. !)

Liane, scandent shrub or shrubby tree. Branches glabrous, grey or reddish brown with numerous lenticels. Petiole 3–5 mm. long, somewhat wrinkled

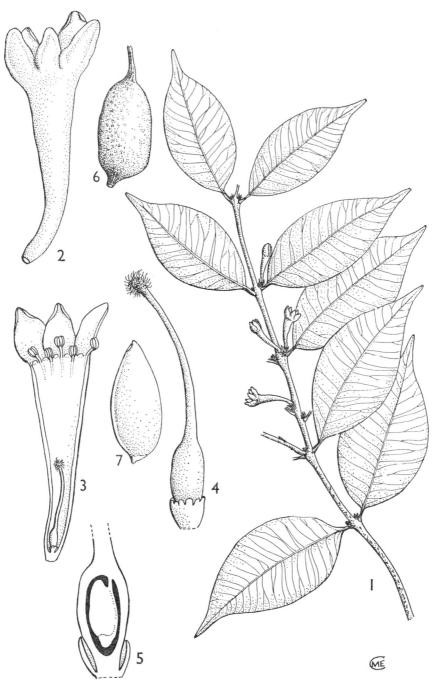

FIG. 3. *CRATEROSIPHON BENIENSE*—**1**, flowering branch, × ⅔; **2**, flower, × 4; **3**, longitudinal section of flower, × 4; **4**, pistil and disc, × 14; **5**, longitudinal section of ovary, × 20; **6**, fruit, × 1; **7**, seed, × 2. 1–5, from *Eggeling* 1720; 6, 7, from *Maitland* 780.

transversely; leaf-blade elliptical to ovate, 35–65(–75) mm. long, 15–35(–40) mm. wide, elongated apex ± 10 mm., glabrous, slightly thickened at the margin. Flowers usually solitary, yellow or greenish yellow; pedicel very short, with ± 1 mm. long ciliate bracteoles. Calyx-tube 8–15 mm. long, 1·3–2 mm. wide at top, glabrous; lobes ovate, 2–4 mm. long, 1–2·5 mm. wide, glabrous; petaloid glands 10 or lacking. Stamens in two whorls ± 0·5 mm. apart; filaments of the stamens opposite the lobes 1 mm. long, the others 0·1–0·2 mm.; anthers 0·5–1 mm. long. Ovary sessile, glabrous; disc lobed, 0·5 mm. high, glabrous; style 4–8 mm. long; stigma papillate. Fruit 23–40 mm. long, 10–16 mm. wide. Seed 15 mm. long, 7 mm. wide. Fig. 3.

UGANDA. Bunyoro District: Budongo Forest, May 1935 (fl.), *Eggeling* 1720!; Mengo District: Kasala Forest Reserve, Dec. 1914 (fl.), *Dummer* 1373!; Masaka District: Sese, Bugoma, June 1926 (fr.), *Maitland* 780!
DISTR. U2, 4; Cameroun, Zaire
HAB. Forest and forest margins; 1200 m.

NOTE. *Maitland* 593 (K!) from Mengo District, Nambigirwa, is probably *C. beniense* but the specimen is sterile. *Maitland* 780 (K!) is cited as *C. scandens* Engl. & Gilg by Burtt Davy in Check-lists Uganda Protectorate: 115 (1935).
   In the specimens from Uganda the calyx-lobes are slightly shorter and more rounded than those from Cameroun and Zaire.

2. C. sp. A

Tall climbing shrub. Branches twining, reddish brown with light lenticels. Leaves opposite or alternate, leathery, dark green above, paler beneath; petiole 3–4 mm. long; blade elliptical, 60–75 mm. long, 25–40 mm. wide, apex ± 5 mm. long, margin slightly undulate. Flowers greenish yellow, clustered; pedicels 4–5 mm., peduncle 6–8 mm. with ciliate bracts. Calyx-tube ± 15 mm. long, 2–3 mm. wide at the top, glabrous outside, slightly hairy inside up to the insertion of the stamens; lobes narrowly ovate, 5–6 mm. long, 1·5–2 mm. wide at the base. Petaloid glands present. Stamens in two whorls, the upper, episepalous ones in the throat of the tube, the second ± 3 mm. lower; filaments 1–1·5 mm. long; anthers 1 mm. long, those of the upper whorl slightly exserted. Ovary glabrous; disc lobed, 0·5–0·8 mm. high; style ± 10 mm. long. Fruit not known.

TANZANIA. Mpanda District: Kabwe R., S. of Pasagulu, 11 Aug. 1959 (fl.), *Harley* 9294!
DISTR. T4; not known elsewhere
HAB. In dry riverine forest; 1460 m.

NOTE. This is very closely allied to *C. scandens* Engl. & Gilg from W. Africa. It differs in having much smaller flowers, slightly hairy inside.

4. PEDDIEA

Harv. in J.B. 2: 265 (1840); H.H.W. Pearson in F.T.A. 6(1): 248 (1910); Domke in Bibl. Bot. 27(111): 124 (1934); Staner in B.J.B.B. 13: 334 (1935); Aymonin in Fl. Cameroun 5: 44 (1966); A. Robyns in F.A.C. Thymelaeaceae: 27 (1975)

Shrubs or small trees with glabrous branches. Leaves alternate, often clustered at the ends of the branches; leaf-blade lanceolate to elliptic, glabrous, midrib raised on both sides. Inflorescence terminal, umbellate or rarely racemose. Flowers 4–5-merous. Calyx-tube cylindric. Petals none. Stamens 8–10 in two whorls on the upper portion of the calyx-tube. Ovary sessile or shortly stipitate, 2-chambered; disc cup-shaped, entire, dentate or lobed, rarely ciliated; style filiform. Fruit a drupe. Seed 1(–2).

A genus with about 10 species in tropical and south-eastern Africa and one (endemic) from Madagascar. Specific limits very uncertain.

Inflorescence racemose   .   .   .   .   .   1. *P. polyantha*
Inflorescence umbellate :
   Pedicel hairy (at top) to puberulous :
     Calyx-tube 6–10 mm. long, slightly hairy to al-
       most glabrous   .   .   .   .   2. *P. montana*
     Calyx-tube 15–20 mm. long, puberulous (later
       nearly glabrous)   .   .   .   .   3. *P. puberula*
   Pedicel glabrous :
     Leaves obtuse or acuminate at base :
       Disc ciliate ; calyx-tube 8–15 mm. long .   .   4. *P. fischeri*
       Disc not ciliate :
         Calyx-tube 20–24 mm. long ; leaves 100–180
           mm. long   .   .   .   .   .   5. *P. lanceolata*
         Calyx-tube 3·5–6 mm. long ; leaves 35–113
           mm. long   .   .   .   .   .   6. *P. rapaneoides*
   Leaves cordate or subcordate   .   .   .   7. *P. subcordata*

See also *P. africana* var. *schliebenii* (p. 15), an insufficiently known taxon.

1. **P. polyantha** *Gilg* in E.J. 30 : 361 (1901) ; H.H.W. Pearson in F.T.A. 6(1) : 250 (1910) ; T.T.C.L. : 610 (1949) ; Archangelsky in Kuprianova, Pollen Morphology : 142, t. 13/8, 18 (1971). Type : Tanzania, Rungwe Mt., *Goetze* 1167 (B, holo. †, K, L, P, iso. !)

Shrub or small tree 2–6 m. high. Branches light brown, smooth. Petiole 3–6 mm. long ; leaf blade lanceolate or elliptic-lanceolate, (70–)92–138(–162) mm. long, (20–)25–38(-43) mm. wide, acuminate, basally cuneate, undulate at the margin, sul coriaceous. Inflorescence racemose, 10–18-flowered ; peduncle 15–35(–62) u m long, slightly hairy. Flowers 4-merous, yellowish green, slightly violet tinged outside ; pedicel 4–6(–10) mm. long, slightly hairy. Calyx-tube 6(–8) mm. long, ± 2 mm. wide in the middle, expanded over the ovary, finely pubescent ; lobes ovate-triangular, 2–2·5 mm. long and wide, ± pubescent on both sides. Anthers 0·5–0·8 mm. long, sessile. Ovary 2–3 mm. long, densely hairy at top ; disc cup-shaped, diminutive ; style 2 mm. long. Mature fruit not seen.

TANZANIA. Iringa District : " Mufu Ukwama ", 6 Sept. 1958 (fl.), *Ede* 16 ! ; Rungwe District : Rungwe Mt., N. slopes, 9 Sept. 1966 (young fr.), *Gillett* 17692 ! & Kiwira Forest Reserve, 8 June 1962 (fl.), *Mgaza* 473 !
DISTR. T7 ; not known elsewhere
HAB. Upland rain-forest ; 1500–2200 m.

2. **P. montana** *Domke* in N.B.G.B. 11 : 670 (1932) ; T.T.C.L. : 610 (1949). Type : Tanzania, Ulanga District, *Schlieben* 2188 (B, holo. †, BM, BR, G, HBG, K, P, S, iso. !)

Shrub 1–2 m. high. Petiole 4–5 mm. long ; leaf-blade lanceolate or ovate-elliptic, 72–105 mm. long, 25–30 mm. wide, acuminate, attenuate to the base, undulate. Inflorescence umbellate, 4–8-flowered ; peduncle 8–15 mm. long ; bracts ± 8 mm. long, slightly ciliate. Flowers 4-merous, yellowish green ; pedicel 8–13 mm., slightly hairy at top. Calyx-tube 6–10 mm. long, glabrous or slightly hairy at base, sometimes with scattered hairs on the lower part ; lobes ovate, 2·5–3 mm. long, 1–1·5 mm. wide, ciliate. Stamens of the upper whorl slightly exserted ; anthers subsessile. Ovary hairy at apex ; disc crenate or lobed, glabrous ; style ± 2 mm. Fruit not seen.

TANZANIA. Ulanga District : SSW. Mahenge, Muhulu Mt., 12 May 1932 (fl.), *Schlieben* 2188 !
DISTR. T6 ; not known elsewhere
HAB. Mist-forest, undergrowth ; 1200 m.

3. **P. puberula** *Domke* in N.B.G.B. 12: 386 (1935); T.T.C.L.: 610 (1949). Type: Tanzania, NW. Uluguru Mts., *Schlieben* 2928 (B, holo. †)

Subshrub to shrub 1–3 m. or small tree 6–8 m. high. Petiole 1–7 mm. long; leaf-blade lanceolate or rarely oblanceolate, 60–120 mm. long, 20–50 mm. wide, acuminate, basally obtuse or subobtuse, subcoriaceous. Inflorescence umbellate, ± 10-flowered; peduncle 15–20 mm. long, puberulous. Flowers 4–5-merous, greenish yellow, paler within, purplish at base; pedicel 2–5 mm. long, puberulous. Calyx-tube 15–20 mm. long, puberulous, later sometimes nearly glabrous; lobes 2·5–4 mm. long, 1·5–3 mm. wide, inner shorter and somewhat broader, ± puberulous on both sides, revolute. Anthers ± 1 mm. long, filaments of same length. Ovary 2 mm. long, densely hairy at apex; disc deeply lobed, glabrous. Young fruits green.

TANZANIA. Morogoro District: NW. side of Uluguru Mts., 8 Nov. 1932 (fl.), *Schlieben* 2928 & S. Uluguru Forest Reserve, Lukwangule Plateau, 17 Mar. 1953 (fl., young fr.), *Drummond & Hemsley* 1644! & above Chenzema towards Lukwangule Plateau, 3 Jan. 1975 (fl.), *Polhill & Wingfield* 4675!
DISTR. **T6**; not known elsewhere
HAB. Upland rain-forest; 1300–2200 m.

4. **P. fischeri** *Engl.*, Hochgebirgsfl. Trop. Afr.: 310 (1892); H.H.W. Pearson in F.T.A. 6(1): 249 (1910); Staner in B.J.B.B. 13: 337 (1935); F.P.N.A. 1: 654, t. 68 (1948); T.T.C.L.: 610 (1949); I.T.U., ed. 2: 425 (1952); K.T.S.: 557 (1961); F.F.N.R.: 272, fig. 48 (1962); Aymonin in Fl. Cameroun 5: 45, t. 6 (1966); Archangelsky in Kuprianova, Pollen Morphology: 142, t. 4/11–12 (1971); J.G. Adam in Mém. Mus. Nat. Hist. Nat., sér. B, Bot., 20: 276, t. 44 (1971); A. Robyns in F.A.C. Thymelaeaceae: 34, t. 6 (1975). Type: Kenya, Rift Valley Province, between Kabaras and Elgeyo [Likaijo], *Fischer* 541 (B, holo. †, K, iso.!)

Shrub or understorey tree up to 6(–10) m. high. Branches glabrous; bark brownish, sometimes striated. Leaves alternate; petiole 2–6 mm. long; leaf-blade lanceolate, oblanceolate or elliptical, cuneate at the base, (30–)65–160(–230) mm. long, (10–)22–50(–68) mm. wide, obtuse to acute, membranous to subcoriaceous. Inflorescence an umbel, (5–)8–12(–26)-flowered; peduncle 6–50 mm. long; bracts lanceolate to ovate, 6–12 mm. long, ciliate, caducous. Flowers 4–5-merous, white or yellowish green, sometimes reddish at base; pedicel 5–20 mm. long. Calyx-tube 8–15 mm. long, glabrous; lobes ovate, 1–2·5 mm. long, 1–2 mm. wide, sparsely to densely pubescent at the apex. Stamens of the upper whorl included or slightly exserted; anthers ± 1 mm., sessile or subsessile. Ovary densely hairy at top, 1·5–3 mm. long; disc 0·3–1·5 mm., sometimes very minute or missing, undulate or irregularly lobed, sparsely or densely ciliate at top; style 1–4 mm. long. Fruit 8–15 mm. long, 6–10 mm. wide, reddish, hairy at top. Fig. 4/1–8.

UGANDA. Kigezi District: Maramagambo Forest, Feb. 1950 (fl.), *Purseglove* 3295!; Busoga District: Lake Victoria, Lolui I., 13 May 1964 (fl., fr.), *G. Jackson* U 65!; Mengo District: Busiro County, Kyiwaga, 22 Sept. 1949 (fl., fr.), *Dawkins* 394!
KENYA. Trans-Nzoia District: Kiminini, Oct. 1968 (fl.), *Tweedie* 3594!; N. Kavirondo District: near Malikisi, Aug. 1959 (fl.), *Tweedie* 1892!; S. Kavirondo District: Bukuria, Sept. 1933 (fl.), *Napier* 2934!
TANZANIA. Mwanza District: Buhindi Forest Reserve, 9 Sept. 1964 (fl.), *Carmichael* 1090!; Morogoro District: N. Uluguru Reserve, above Morningside Hotel, 7 Feb. 1944 (fl.), *Wigg* in *F.H.* 1474!; Iringa District: Irundi, 9 Nov. 1955 (fl.), *Benedicto* 79!
DISTR. **U2–4**; **K3–5**; **T1–3**, 6, 7; Guinée to Cameroun, Zaire, Rwanda, Burundi, Sudan, Malawi, Zambia, Angola
HAB. Forest understorey, margins, associated bushland and thickets, sometimes along rivers; 950–2400 (?–3000) m.

SYN. *P. zenkeri* Gilg in E.J. 19: 256 (1894); H.H.W. Pearson in F.T.A. 6(1): 252 (1910). Type: Cameroun, Yaoundé, *Zenker* 242 (B, holo.†)
     P. longipedicellata Gilg in E.J. 19: 256 (1894); De Wild. in Ann. Mus. Congo, Bot., sér. 4, 1, t. 26/8 (1903); H.H.W. Pearson in F.T.A. 6(1): 250 (1910). Type: Malawi, *Buchanan* 536 (B, holo.†, K, iso.!)

FIG. 4. *PEDDIEA FISCHERI*—**1,** flowering branch, × ⅔; **2,** flower, × 4; **3,** longitudinal section of flower, × 4; **4,** pistil and disc, × 10; **5,** longitudinal section of ovary and disc, × 14; **6,** young fruiting branchlet, × ⅔; **7,** fruit, × 2; **8,** seed, × 4. *P. SUBCORDATA*—**9,** leaf, × ⅔. 1–6, from *Greenway* 9603; 7, 8, from *Semsei* 2199; 9, from *Pócs* 6596/B.

P. longiflora Engl. & Gilg in E.J. 19: 257 (1894); H.H.W. Pearson in F.T.A.
6(1): 249 (1910). Types: Togo, Bismarckburg, *Büttner* 220 & between Jegge
and Konkoa, *Büttner* 481 & Assuma, *Büttner* 287 (all B, syn.†)

P. volkensii Gilg in E.J. 19, Beibl. 47: 41 (1894); H.H.W. Pearson in F.T.A.
6(1): 251 (1910); T.S.K., ed. 2: 19 (1936); T.T.C.L.: 611 (1949); K.T.S.:
557 (1961); Archangelsky in Kuprianova, Pollen Morphology: 142, t. 13/10,
19 (1971). Type: Tanzania, Kilimanjaro, Marangu, *Volkens* 1283 (B, holo.†,
G, K, iso.!)

P. longipedicellata Gilg var. *multiflora* De Wild. in Ann. Mus. Congo, Bot., sér. 4,
1: 94 (1903), t. 26/1–7 (1902); H.H.W. Pearson in F.T.A. 6(1): 250 (1910).
Type: Zaire, Shaba, Lukafu, *Verdick* 565 (BR, holo.!)

P. cyathulata H.H.W. Pearson in F.T.A. 6(1): 251 (1910) & in K.B. 1910:
341 (1910). Type: Malawi, Mt. Malosa, *Whyte* (K, holo.!)

P. batesii S. Moore in J.B. 57: 118 (1919). Type: Cameroun, *Bates* 1035*
(BM, holo.!)

P. potamophila Gilg in V.E. 3(2): 629 (1921), *nom. in clavi*

P. mildbraedii Engl. & Gilg in V.E. 3(2): 629 (1921), *nom. in clavi*

P. multiflora (De Wild.) Engl. in V.E. 3(2): 629 (1921)

?*P. brachypoda* Gilg & Ledermann in V.E. 3(2): 629 (1921), *nom. in clavi*

P. arthuri Th. C.E. Fries in N.B.G.B. 8: 422 (1923). Type: Kenya, Meru,
R. & Th. C.E. Fries 1613 (UPS, holo.!, K, iso.)

P. fischeri Engl. var. *coriacea* Domke in N.B.G.B. 11: 671 (1932); T.T.C.L.:
610 (1949). Types: Tanzania, Njombe District, Lupembe, N. Ruhudji R.,
Schlieben 611 (B, syn.†, G, K, P, S, iso.!) & 659 (B, syn.†, G, iso.!)

Note. This is a very widespread species but the distribution does not seem correlated
with any geographical variation. *P. fischeri* is a very variable species that has been
described under many different names. With more material available showing that
several species have been separated by insignificant or quantitative characters the
synonymy of *P. fischeri* has gradually increased.

The only difference between *P. volkensii* and *P. fischeri* is the attenuated rather
than shortly rounded base of the calyx-tube; I do not consider *P. volkensii* can be
maintained as a distinct species.

5. **P. lanceolata** *Domke* in N.B.G.B. 11 : 670 (1932); T.T.C.L.: 610 (1949).
Type: Tanzania, Ulanga District, *Schlieben* 1908 (B, holo. †, B, G, HBG, K,
P, S, iso.!)

Shrub about 1–2 m. high. Petiole 3–4 mm. long; leaf-blade lanceolate or
oblanceolate, (100–)120–150(–180) mm. long, (15–)22–30(–35) mm. wide, narrowed at base and tip. Inflorescence a terminal umbel, 8–15(–25)-flowered;
peduncle 8–20 mm. long; bracts 5–12 mm. long, 1·5–2·5 mm. wide, glabrous
or slightly ciliate. Flowers 4–5-merous, yellowish green; pedicel 2–4 mm. long.
Calyx-tube 20–24 mm. long, 1–2 mm. wide, glabrous, striate; lobes ovate,
2·5–4 mm. long, 1–2 mm. wide, the interior a little smaller, the exterior
pubescent at the margins. Anthers subsessile, ± 1·5 mm. long. Ovary 2 mm.
long, glabrous; disc 0·5–1·0 mm., crenulate or lobed, glabrous; style ± 5 mm.
long. Fruit not seen.

Tanzania. Ulanga District: 35 km. S. of Mahenge, near Sali, 18 Mar. 1932 (fl.),
Schlieben 1908!
Distr. **T6**; not known elsewhere
Hab. Mist-forest by stream banks; 1000–1100 m.

6. **P. rapaneoides** *Gilg* in V.E. 3(2): 629 (1921) *in clavi*; Staner in B.J.B.B.
13: 336 (1935); F.P.N.A. 1: 653 (1948); A. Robyns in B.J.B.B. 34: 393, fig.
29/A (1964) & in F.A.C. Thymelaeaceae: 28, t. 5/E–G (1975). Type: Rwanda,
Rugege, *Mildbraed* 1016 (B, holo. †, BR, fragm.!)

Much-branched shrub or tree up to 10(–15) m. high. Petiole 2–6 mm. long;
leaf-blade narrowly to broadly elliptic or lanceolate, (35–)50–92(–113) mm.
long, (15–)23–37(–46) mm. wide, obtuse to acute, basally cuneate, subcoriaceous. Inflorescence umbellate, 8–26-flowered; peduncle 10–40 mm.

---

* For some reason the collector's number on the label of the type specimen has been
altered from 1035 to 1036.

long; bracts lanceolate, 6–8 mm. long, ± 3 mm. wide, ciliate. Flowers
4(–5)-merous, greenish yellow; pedicel 6–13 mm. long, glabrous. Calyx-tube
3·5–6·0 mm. long, glabrous; lobes ovate, 1·0–2·3 mm. long, 1·0–2·5 mm. wide,
pubescent at top, reflexed. Anthers subsessile, episepalous slightly exserted.
Ovary ± 2 mm. long, 1·3–1·5 mm. wide, glabrous or with a few hairs at top;
disc ± 1 mm., irregularly lobed, glabrous; style 1–2 mm. long. Fruit 8–11
mm. long, 6·5–8 mm. wide, purple, glabrous.

UGANDA. Kigezi District: Mt. Mgahinga–Sabinio saddle, 21 July 1960 (fl., fr.), *Schaller*
300! & Luhizha, Mar. 1947 (fl., fr.), *Purseglove* 2365! & Mafuga Forest, Sept. 1947
(fl.), *Dale u.* 504!
DISTR. U2; eastern Zaire, Rwanda, Burundi
HAB. Upland rain-forest, sometimes in bamboo zone; 1200–2800 m.

SYN. *P. bambuseti* Gilg in V.E. 3(2): 629 (1921) *in clavi.* Type: Zaire, Sabinio–
Mgahinga saddle, *Mildbraed* 1679 (B, holo.†, BR, fragm.!)

NOTE. In I.T.U., ed. 2: 426 (1952), *Purseglove* 2365 is cited as " *P. rhododendroides* ",
a typographic error.

7. **P. subcordata** *Domke* in N.B.G.B. 12: 387 (1935); T.T.C.L.: 610 (1949).
Type: Tanzania, NW. Uluguru Mts., *Schlieben* 3092 (B, holo. †, B, BM, G, P,
S, iso.!)

Shrub or small tree up to 2 m. high. Branches slender, glabrous. Petiole
1–5 mm. long; leaf-blade lanceolate or rarely oblanceolate or obovate,
(120–)146–214(–234) mm. long, (39–)44–60(–68) mm. wide, slightly acute,
basally cordate or subcordate, membranous, margin ± wavy. Inflorescence
a terminal umbel, 20–34(–55)-flowered; peduncle 15–30 mm. long. Flowers
4-merous, greenish yellow; pedicel 3–5 mm. long. Calyx-tube 8–16 mm.
long, usually striate, glabrous; lobes ovate, 2 mm. long, 1·2–2·0 mm. wide,
slightly puberulous on the inside of the margin. Anthers 0·8–1·0 mm., subsessile. Ovary 2–3 mm. long, glabrous; disc 0·3–0·5 mm., irregularly lobed,
glabrous; style 4–8 mm. Fruit blackish. Seed 7–9 mm. long, 4–5 mm. wide.
Fig. 4/9, p. 13.

TANZANIA. Morogoro District: S. Nguru Mts., Ruhamba Peak, 2 Apr. 1953 (fr.),
Drummond & Hemsley 1982! & Uluguru Mts., Bondwa Peak, Jan. 1953 (fl.), *Eggeling*
6471! & Bunduki Forest Reserve, Mar. 1953 (fl.), *Paulo* 54!
DISTR. T6; known only from the Nguru, Ukaguru and Uluguru Mts.
HAB. Upland rain-forest; 1000–2000 m.

Insufficiently known taxon

P. africana *Harv.* var. **schliebenii** *Domke* in N.B.G.B. 12: 388 (1935);
T.T.C.L.: 610 (1949). Type: Tanzania, Uluguru Mts., *Schlieben* 2928a (B,
holo. †)

Shrub up to 2 m. or a tree 4–8 m. high. Leaf-blade lanceolate to elliptic,
25–120 mm. long, 20–50 mm. wide. Inflorescence an umbel, 6–12-flowered;
peduncle 10–20 mm. long. Flowers 4–5-merous, greenish yellow; pedicel
4–9 mm. long. Calyx-tube 8–12 mm. long, ribbed, glabrous or rarely slightly
hairy; lobes ovate, 2–3·5 mm. long, ± 2 mm. wide, glabrous or with slightly
puberulous margins. Filaments 1 mm. long; anthers 1 mm. long. Ovary
2–3 mm. long, glabrous or slightly to densely hairy at top; disc ± 1 mm.,
sometimes lobed, glabrous; style 3 mm. long. Fruit black, 10–12 mm. long,
8–10 mm. wide, glabrous or rarely hairy at top.

TANZANIA. Morogoro District: Uluguru Mts., 8 Nov. 1932 (fl.), *Schlieben* 2928a
DISTR. T6; not known elsewhere
HAB. Rain-forest; 1300 m.

NOTE.   The variety *schliebenii* only differs from *P. africana* in its more slender pedicels and calyx-tubes.   The variety is also said to have a non-fimbriate disc.   There is a great variation of these characters within *P. africana* proper, which occurs in Rhodesia and Mozambique south to NE. Cape Province of South Africa.   In the absence of authentic material no decision can be made on the identity of this variety.   Material from the type locality is needed.   *P. africana* var. *schliebenii* and *P. puberula* are both based on material collected by *Schlieben* at the same locality.

5. DAIS

L., Sp. Pl., ed. 2, 1 : 556 (1762) & Gen. Pl., ed. 6 : 215 (1764); Domke in Bibl. Bot. 27(111): 133 (1934)

Deciduous, many-branched shrubs or trees.   Branches dark or greyish brown, striate, glabrous.   Leaves often at the ends of the branches, opposite or alternate, petiolate; blade glabrous, with a slightly bluish tinge above, light green beneath, midrib and pinnate lateral veins yellow beneath, slightly raised, lateral veins curve before reaching the margin, indistinctly reticulate. Inflorescence a dense, peduncled, terminal head.   Involucral bracts rigid, persistent.   Calyx-tube cylindric, often slightly curved, circumscissile above the ovary; lobes 5 (rarely 4), outer slightly larger than the inner.   Petals lacking.   Stamens 10, 2-seriate.   Ovary 1-chambered; disc cupular, irregularly lobed; style lateral, filiform; stigma capitate, papillose.

Two species, one in tropical and south-eastern Africa and one from Madagascar.

D. cotinifolia *L.*, Sp. Pl., ed. 2, 1 : 556 (1762); Meisner in DC., Prodr. 14 : 528 (1857) incl. vars.; Verdc. in K.B. 11 : 453 (1957); Peterson in Bot. Notis. 111 : 631 (1958); Archangelsky in Bot. Zhurn. 51, t. 4/10 (1966) & in Kuprianova, Pollen Morphology: 187, t. 12/12 (1971).   Type: *Herb. Linnaeus* 554.1 (LINN!)

Shrubs or trees up to 8(–15) m.   Petiole (2–)4–6 mm.; leaf-blade broadly lanceolate to elliptical, (25–)30–90(–150) mm. long, (15–)20–50(–65) mm. wide, acute to obtuse, subcoriaceous.   Heads with 20–60 flowers, peduncle up to 8 cm., receptacle flat, with bristles around the flower-insertions.   Bracts ovate to almost orbicular, chestnut brown, becoming black, coriaceous, 8–16 mm. long, 5–14 mm. wide, outer ones largest, ± hairy at the margins especially at apex, inner smaller, hairy beneath in the middle part.   Flowers pale lilac, pink or white.   Calyx-tube silky villous outside, less densely inside, 10–30 mm. long; lobes narrowly ovate, unequal, 4–8(–10) mm. long, 1–2(–3) mm. wide, puberulous externally, less so internally.   Stamens included or the upper exserted; filaments 0·5–1·5 mm. long; anthers 1–1·5 mm. long.   Ovary densely pilose at the top, otherwise glabrous; disc ± 1 mm.; stigma in most cases exserted.   Fruit dry, enclosed in the base of the calyx-tube.   Seed black, crustaceous.   Fig. 5.

TANZANIA.   Iringa District: Johns Corner, 12 Mar. 1962 (fl.), *Polhill & Paulo* 1730! & Mufindi, Irundi, Jan. 1961 (fl.), *Procter* 1701!; Njombe District: Njombe–Uwemba road, Mpala Forest, May 1953 (fl.), *Eggeling* 6549!
DISTR.   T7; Malawi, Rhodesia, South Africa (Transvaal to King Williamstown District), Swaziland, Lesotho
HAB.   Margins of upland forests and associated bushland, river banks; 1700–2300 m.

SYN.   *D. laurifolia* Jacq., Coll. Bot. 1 : 146 (1787) & Ic. Pl. Rar. 1 : 8, t. 77 (1787), *non* Blanco (1837).   Type: Jacquin, Icones Plantarum Rariorum 1, t. 77 (1787), probably no specimen was kept of the plant cultivated at Schoenbrunn and described and figured by Jacquin
        Lasiosiphon grandifolius Gilli in Ann. Naturhist. Mus. Wien 74: 450, t. 7/2 (1970).   Type: Tanzania, Njombe District, Uwemba, *Gilli* 358 (W, holo.!)
        Gnidia grandifolia (Gilli) Gilli in Ann. Naturhist. Mus. Wien 74: 451 (1970), pro syn.

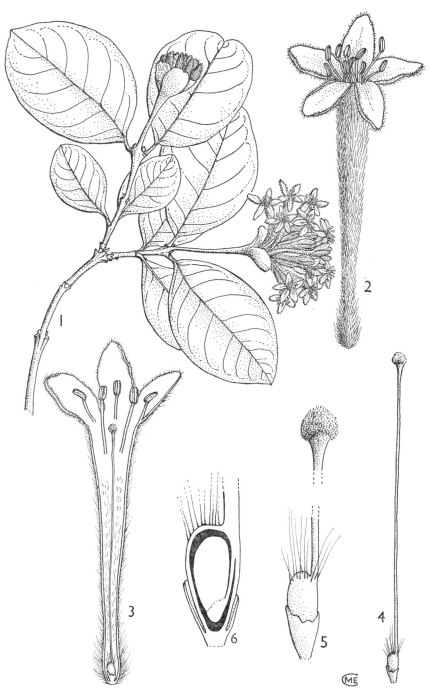

FIG. 5. *DAIS COTINIFOLIA*—**1,** flowering branch, × ⅔; **2,** flower, × 3; **3,** longitudinal section of flower, × 3; **4,** pistil and disc, × 4; **5,** details of same, × 10; **6,** longitudinal section of ovary and disc, × 16. All from *Richards* 14072.

NOTE.   The species has been introduced from South Africa as an ornamental shrub in
the highlands of Kenya and Tanzania, e.g. Nairobi, *Greenway* 14084, see Jex-Blake,
Gardening in E. Afr., ed. 4: 110 (1957).

6. GNIDIA

L., Sp.Pl.: 358 (1753) & Gen. Pl., ed. 5: 168 (1754); Staner in B.J.B.B. 13: 340
      (1935); Peterson in Bot. Notis. 112: 465 (1959); Aymonin in Fl. Gabon 11: 88
      (1966) & in Fl. Cameroun 5: 49 (1966); Gastaldo in Webbia 24: 338 (1969);
      A. Robyns in F.A.C. Thymelaeaceae: 36 (1975)

Lasiosiphon Fresen. in Flora 21: 602 (1838); H.H.W. Pearson in F.T.A. 6(1):
      227 (1910); Domke in Bibl. Bot. 27(111): 133 (1934)
Arthrosolen C.A. Mey. in Bull. Acad. Imp. St. Petersb. 1: 359 (1843); H.H.W.
      Pearson in F.T.A. 6(1): 234 (1910)
Englerodaphne Gilg in E.J. 19: 274 (1894); H.H.W. Pearson in F.T.A. 6(1):
      237 (1910)

Perennial herbs, shrubs or trees. Bark smooth or rough, with or without
lenticels. Leaves alternate, rarely opposite, often ericoid. Inflorescence spi-
cate or ebracteate, few-flowered fascicles or bracteate, few–many-flowered
terminal or axillary heads. Flowers ♀, 4–5-merous, sessile or pedicelled.
Calyx-tube cylindric, glabrous or pubescent, usually articulated above the
ovary, with the upper part falling off after flowering; lobes imbricate, always
shorter than the tube, erect, spreading or reflexed, coloured. Petals alternating with the calyx-lobes, entire or ± deeply divided, membranous or fleshy,
often scaly or glandular, sometimes missing. Stamens in one whorl (not in East
Africa) or two whorls in the throat of the calyx-tube, those of the upper whorl
often partly exserted, opposite the calyx-lobes; anthers sessile or with very
short filaments, sometimes heteromorphic. Ovary sessile or shortly stipitate,
1-locular, glabrous or pubescent; hypogynous disc membranous, cup-shaped
or lobed, sometimes minute or lacking; style filiform, lateral; stigma capitate
or club-shaped, papillate. Fruit dry, small, enclosed by the persistent base of
the calyx-tube. Seed with scanty or no endosperm.

The largest genus within the family, about 140 species, the major part in tropical and
southern Africa but extending to Arabia (one species) and western India and Sri Lanka
(two species). About 20 species endemic in Madagascar.
      The number of calyx-lobes and the presence or absence of petals (petaloid scales) are
the only generic points of distinction between *Gnidia*, *Lasiosiphon* and *Arthrosolen*. One
and the same individual can have both 4- and 5-merous flowers. The petals can be well
developed or fully aborted within one and the same species. Considering the great
variation in these characters and in agreement with the opinion of most recent authors
Lasiosiphon and *Arthrosolen* as well as *Englerodaphne* are here treated as synonyms of
Gnidia. See also discussion by Peterson in Bot. Notis. 112: 465–467 (1959).

Flower-parts mostly in fours:
   Flowers either in terminal ebracteate clusters (fig.
         6/1) or 1–4 terminal and in the axils subtended
         by leaf-like bracts; petals (6–)8:
      Flowers (2–)4–6(–8) in terminal clusters; leaves
            opposite, ovate to round, glabrous, with
            pinnate nerves and reticulate venation   .   1. *G. subcordata*
      Flowers 1–4 in the axils; leaves alternate, linear-
            lanceolate, hairy with faint ascending nerves   2. *G. fastigiata*
   Flowers in heads subtended by an involucre of
         clearly differentiated bracts; petals 0–4:
      Heads 10–60-flowered, with a single series of in-
            volucral bracts; stem and leaves glabrous or
            virtually so:
         Calyx-lobes white to dull yellow or red, obtuse
               or rounded; stamens all similar; petals

small but generally present (except sometimes in *G. microcephala*) :

Bracts (fig. 6/3a) ciliate, linear to ovate, acuminate, papery; calyx-tube generally
glabrous below the articulation, sometimes sparsely hairy :

Heads terminal and axillary; stems much
branched, to 1 m. . . . . . 3. *G. apiculata*

Heads terminal; stems not or little
branched, to 30 cm. . . . 4. *G. microcephala*

Bracts not ciliate :

Calyx-tube below the articulation densely
covered with long hairs; bracts 1-
nerved from the base, lanceolate to
ovate-lanceolate, acute to acuminate,
firmly textured (fig. 6/6) . . . 6. *G. stenophylla*

Calyx-tube below the articulation glabrous
or sometimes in *G. goetzeana* shortly
hairy; bracts several-nerved from the
base, ovate-lanceolate to suborbicular :

Petals (fig. 6/5c) filiform with a small
head or glandular; bracts (fig. 6/5a)
ovate-lanceolate to ovate, acute or
acuminate, firmly textured, deciduous in fruit; stems 1(–several)
from a tap-root, to 1·5 m., well
branched . . . . . 5. *G. goetzeana*

Petals (fig. 6/7c) linear to obovate or
spathulate, variably developed;
bracts (fig. 6/7a) elliptic-oblong,
ovate or suborbicular, acute to
rounded, ± papery, persistent;
stems generally numerous from a
woody rootstock :

Heads 15–25(–40)-flowered; petals
obovate or spathulate; disc minute . . . . . 7. *G. involucrata*

Heads 30–60-flowered; petals linear
to narrowly elliptic; disc none . 8. *G. usafuae*

Calyx-lobes (fig. 7/4, p. 27) brilliant yellow to
red, usually acuminate, occasionally blunt;
stamens slightly dimorphic, the upper ones
somewhat longer; petals 0 . . . 9. *G. chrysantha*

Heads 50–100-flowered, composite, made up of
numerous small 2–6(–12)-flowered heads,
each with 4 bracteoles; branches and leaves
hairy . . . . 10. *G. mollis*

Flower-parts mostly in fives :

Perennial suffrutescent herb up to 60 cm. high . 11. *G. kraussiana*

Much-branched shrub or tree :

Bracts with ciliate margins, otherwise glabrous 12. *G. eminii*

Bracts pubescent or tomentose on the outside :

Inflorescence 6–12-flowered . . . 13. *G. latifolia*

Inflorescence 20–70-flowered :

Leaves densely pubescent, sometimes par-
tially glabrescent with age . . 14. *G. lamprantha*

Leaves glaucous, glabrous or glabrescent . 15. *G. glauca*

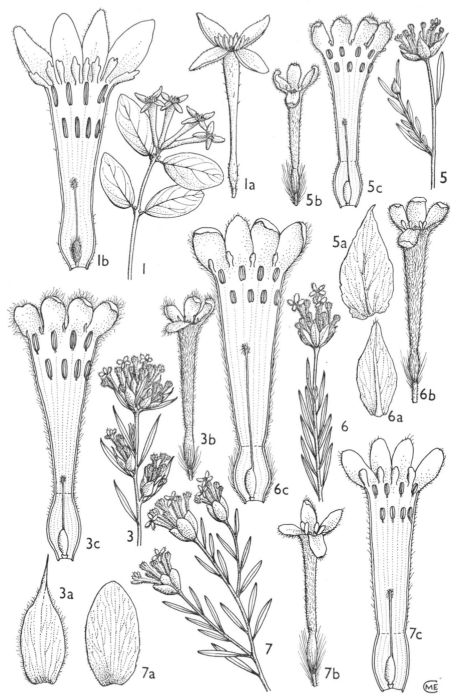

Fig. 6. Flowering branchlets (× 1), bracts, flowers and longitudinal sections of flowers of *Gnidia* (species numbered as in text). **1**, *G. subcordata*, 1a × 3, 1b × 4; **3**, *G. apiculata*, 3a, 3b × 4, 3c × 6; **5**, *G. goetzeana*, 5a, 5b × 4, 5c × 6; **6**, *G. stenophylla*, 6a, 6b × 4, 6c × 6; **7**, *G. involucrata*, 7a, 7b × 4, 7c × 6. 1, from *Eggeling* 6819; 3, from *Polhill & Paulo* 1070; 5, from *Milne-Redhead & Taylor* 9643; 6, from *Gillett* 14104; 7, from *Milne-Redhead & Taylor* 9257.

1. **G. subcordata** *Meisner* in Linnaea 14: 430 (1840) & in DC., Prodr. 14: 586 (1857); U.K.W.F.: 161 (1974). Type: South Africa, Tembuland, Mqanduli, near Morley, *Drège* 4670 (NY-MEISN, lecto.!, P, iso.!)

Much-branched shrub up to 3·5 m. high. Branches slender, glabrous; bark dull or purplish grey. Leaves opposite, with petiole 0·5–1 mm. long; leaf-blade ovate or suborbicular, apex acute, rounded or truncate, base rounded or truncate, sometimes subcordate, 10–25(–32) mm. long, 5–13(–16) mm. wide, membranous, green above, glaucous-green beneath, glabrous or with occasional hairs beneath along the midrib. Inflorescence a terminal (2–)4–6(–8)-flowered, ebracteate fascicle. Flowers greenish white, white or cream, 4-merous, shortly pedicelled. Calyx-tube 9–12 mm. long, constricted below the middle, glabrous or with few silky hairs; lobes oblong, spreading during flowering, 3·5–5 mm. long, ± 2 mm. wide, glabrous or sometimes with a few hairs. Petals 8 somewhat fleshy lobes, linear-spathulate, emarginate or toothed at the apex, 1–3 mm. long, ± 1 mm. wide. Stamens subsessile, anthers 1–1·5 mm. long, upper whorl slightly exserted. Ovary shortly stalked, densely hirsute; disc none; style 4–6 mm. long. Fig. 6/1.

UGANDA. Karamoja District: near Kamion, June 1942 (fl.), *Dale* u.256! & Timu Forest, 7 Apr. 1951 (fl.), *D. Bacon* 10! & Mar. 1960 (fl.), *J. Wilson* 926!
KENYA. Turkana District: Murua Nysigar [Morussigar], 17 Feb. 1965 (fl.), *Newbould* 7250!; Nairobi, at edge of Thika Road, 19 Nov. 1950 (fl.), *Verdcourt* 382!; Masai District: 19 km. on Narok–Olokurto road, Orengitok, 17 May 1961 (fl.), *Glover, Gwynne & Samuel* 1252!
TANZANIA. Kilimanjaro, near Rongai, Jan. 1953 (fl.), *Eggeling* 6484!; Mbulu District: 8 km. on Babati–Singida road, 6 May 1962 (fl.), *Polhill & Paulo* 2356!; Singida District: 13 km. from Mkalama, Oct. 1935 (fl.), *B.D. Burtt* 5287!
DISTR. U1; K2–7; T2, 3, 5; Sudan, South Africa (Natal to King Williams Town)
HAB. Upland dry evergreen forest and associated bushland, *Acacia* woodland and wooded grassland; 1400–2400 m.

SYN. *Englerodaphne leiosiphon* Gilg in E.J. 19: 274 (1894) & 18: 504, fig. 7/A–E (1894) & in E. & P. Pf. III. 6a: 230, fig. 81/A–E (1894); H.H.W.Pearson in F.T.A. 6(1): 238 (1910); T.S.K., ed. 2: 18 (1936); F.P.S. 1: 148 (1950); Archangelsky in Bot. Zhurn. 51, t. 3/15–16 (1966) & in Kuprianova, Pollen Morphology: 193, t. 14/1–2 (1971). Type: Kenya, Kitui, *Hildebrandt* 2751 (B, holo.†, G, K, L, P, iso.!)
E. subcordata (Meisner) Engl., V.E. 3(2): 640, fig. 280/A–E (1921); T.T.C.L.: 607 (1949); K.T.S.: 555, fig. 102 (1961)
Gnidia leiosiphon (Gilg) Domke in Bibl. Bot. 111: 90 (1934)

2. **G. fastigiata** *Rendle* in Trans. Linn. Soc. Bot., ser. 2, 4:41 (1894); H.H.W. Pearson in F.T.A. 6(1): 222 (1910); T.T.C.L.: 607 (1949); Peterson in Bot. Notis. 111: 624 (1958); U.K.W.F.: 161 (1974). Type: Malawi, Mulanje [Mlanje] Mt., *Whyte* (K, holo.!, BM, iso.!)

Subshrub up to 40 cm. high. Branches puberulous or hirsute when young, later glabrescent. Leaves crowded on the upper part of the branches, sessile; leaf-blade linear-lanceolate, (6–)10–12(–16) mm. long, 1–2(–3) mm. wide, rigid, somewhat wrinkled, sparsely silky-hairy to densely hirsute on the lower surface, sparsely hairy on the upper, later glabrescent. Inflorescences terminal or axillary clusters of 2–4 flowers (sometimes only one flower), sessile or very shortly stalked; involucral bracts 2–4, leaf-like, 7–8 mm. long, 1–1·2 mm. wide, silky-hairy. Flowers pale blue, whitish or yellow, 4-merous; pedicel very short, with short brush of hair. Calyx-tube 4–8 mm. long, densely villous; lobes ovate, 2–2·5 mm. long, 1 mm. wide, villous outside, sometimes hairy at top and along middle inside. Petals (6–)8, linear, membranous, 0·5–1 mm. long. Stamens subsessile, anthers 0·5 mm. long; a tuft of hairs behind the anthers of the upper row. Ovary ± 1 mm. long, hairy at top; disc cupular; style 3–4 mm. long, lateral; stigma club-shaped, papillate. Seed 2 mm. long, 1–1·2 mm. wide.

UGANDA.   Karamoja District: Mt. Moroto, May 1963 (fl., fr.), *J. Wilson* 1388!
KENYA.   Turkana District: Karasuk, Mt. Lorosuk, 15 Aug. 1965 (fl., fr.), *J. Wilson* 1672!; Elgeyo District: Marakwet Hill, Feb. 1934 (fl., fr.), *Dale* in *F.D.* 3175!; Ravine District: Timboroa, 29 June 1956 (fl., fr.), *Irwin* 284!
TANZANIA.   N. slopes of Kilimanjaro, above Rongai, 1 Dec. 1932 (fl.), *C.G. Rogers* 169!; Mpanda District: Mahali Mts., Sisaga, 26 Aug. 1958 (fl.), *Newbould & Jefford* 1775!; Rungwe District: Kiwira R., upper fishing camp, 25 Oct. 1947 (fl.), *Greenway & Brenan* 8277!
DISTR.   **U**1; **K**2, 3, 5; **T**2–4, 7; Malawi, Rhodesia
HAB.   Mountain grasslands, rocky hillsides; 1000–3000 m.

SYN.   *Struthiola holstii* Engl. in E.J. 17: 165 (1893), *nom. nudum*
        Gnidia holstii Engl. & Gilg in E.J. 19: 257 (1894); Gilg in P.O.A. C, t. 32/K–M (1895). Type: Tanzania, Lushoto District, Mlalo, *Holst* 252 (B, holo.†)
        G. holstii Engl. & Gilg var. *kilimandscharica* Gilg in P.O.A. C: 283 (1895). Type: Tanzania, Moshi District, above Usseri, *Volkens* 1904 (B, holo.†, BM, E, G, K, iso.!)
        G. fastigiata Rendle var. *hirsuta* H.H.W. Pearson in F.T.A. 6(1): 223 (1910); T.T.C.L.: 608 (1949). Types: Kenya, Nandi, *Scott Elliot* 6966 (BM, K, syn.!) & Upper Mau Plateau, *Whyte* (K, syn.!) & Tanzania, Njombe District, Ussangu, Pikurugwe, *Goetze* 1258 (BM, E, G, K, P, syn.!) & Malawi, between Mpata and Tanganyika Plateau, *Whyte* (K, syn.!) & Nyika Plateau, *Whyte* (K, syn.!) & Tuchila Plateau, *Purves* 112 (K, syn.!)

3.   **G. apiculata** (*Oliv.*) *Gilg* in E.J. 19: 263 (1894), pro parte, excl. saltem specim. *Dewèvre* (=*Descamps* 67), *Buchanan* 89 & *Whyte*; H.H.W. Pearson in F.T.A. 6(1): 220 (1910); Fries in N.B.G.B. 8: 421 (1923); T.T.C.L.: 607 (1949); F.P.S. 1: 150 (1950); U.K.W.F.: 161 (1974); A. Robyns in F.A.C. Thymelaeaceae: 45, fig. 4 (1975). Type: Sudan, Madi, *Grant* (K, holo.!)

Erect, much branched shrub or undershrub up to 1 m. high. Branches glabrous; bark reddish brown. Leaves sessile or subsessile; leaf-blade linear or narrowly elliptic, (4–)10–24(–32) mm. long, (0·5–)1–5(–8) mm. wide, rigid, glabrous. Inflorescence capitate, terminal and axillary, 10–25-flowered, sessile or shortly peduncled. Bracts 8–14, narrowly linear to lanceolate, 6–10 mm. long, 2·5–4 mm. wide, papery, densely ciliate, inner ± puberulous, especially along the midrib, outer usually glabrous, brownish, persistent. Flowers yellow or greenish yellow, 4-merous; pedicel 1–2 mm., with white hairs, the upper ones surrounding the ovary. Calyx-tube 7–12 mm. long, usually glabrous but sometimes sparsely hairy below the articulation, densely pubescent above the articulation; lobes oblong, rounded at the apex, 1–1·5 mm. long, 0·8–1 mm. wide, pubescent on the outside. Petals minute, spathulate or filiform, 0·3–0·5 mm. long. Stamens sessile or subsessile; anthers 0·5–0·8 mm. long. Ovary 1·5–2·5 mm. long, shortly stipitate, glabrous; disc minute or lacking; style 2–3 mm. long, excentric; stigma papillate. Mature fruits not seen. Fig. 6/3, p. 20.

UGANDA.   W. Nile District: below Metu Rest Camp, 12 Sept. 1953 (fl.), *Chancellor* 248! & Mt. Otzi, Oct. 1959 (fl.), *E.M. Scott* in *E.A.H.* 11797! (both are f. *pyramidalis*)
KENYA.   Fort Hall District: Thika, W. of Blue Posts Hotel, 15 Jan. 1967 (fl.), *Faden* 6736!; Embu, 5 Feb. 1960 (fl.), *Rauh* K 328!; Kitui Hills, Lower Ukamba, Jan. 1937 (fl.), *Gardner* 3603!
TANZANIA.   Mbulu District: Pienaars Heights, between Babati and Bereko, 6 Jan. 1962 (fl.), *Polhill & Paulo* 1070!; E. Kilimanjaro, Rombo, May 1927 (fl.), *Haarer* 658!; Morogoro District: N. Uluguru Reserve, above Morningside, May 1953 (fl.), *Semsei* 1221! (= f. *pyramidalis*)
DISTR.   **U**1; **K**4, 5; **T**2, 3, 5–7; Cameroun, Gabon, Zaire, Chad, Central African Empire, Sudan
HAB.   Grasslands, dry hillsides, on red soil; 150–1700 m.

SYN.   *G. involucrata* A. Rich. var. *apiculata* Oliv. in Trans. Linn. Soc. 29: 143, t. 91 (1875)

NOTE.   Plants coming up after fires resemble *G. microcephala* and, indeed, the distinction between these two species is not very sharp where the ranges approximate in Tanzania, though plants from further north and south are markedly different.

A form with more narrow leaves and more elongate and slender branches has been described from western and central Africa as forma *pyramidalis* by G. Aymonin in Fl. Gabon 11: 95, t. 14/6 (1966) & in Fl. Cameroun 5: 74, t. 7/6, 8/1–8 (1966); type: Central African Empire, Oubangui, *Tisserant* 307 (P, holo !, BM, P, iso. !). Undoubtedly some specimens from Uganda and Tanzania belong to this form.

G. gossweileri S. Moore from Angola is very closely related to *G. apiculata* and may prove to be conspecific.

4. **G. microcephala** *Meisner* in DC., Prodr. 14: 589 (1857); H.H.W. Pearson in F.T.A. 6(1): 225 (1910); Peterson in Bot. Notis. 111: 625 (1958) & in Bol. Soc. Brot., sér. 2, 33: 212 (1959). Type: South Africa, Transvaal, Magaliesberg, *Zeyher* 1492 (G, P, S, SAM, iso. !)

Herbaceous or somewhat woody perennial up to 30 cm. high from a woody rhizome. Stems numerous, glabrous, loosely branched. Leaves sessile or subsessile; leaf-blade linear, lanceolate to narrowly obovate, acuminate or acute, (6–)10–18(–22) mm. long, 0·5–3 mm. wide, margins sometimes inrolled, glabrous. Inflorescences up to 30-flowered terminal heads, sometimes overtopped by the uppermost leaves. Bracts (4–)6–8(–10), lanceolate to broadly ovate, acuminate to acute, 6–10 mm. long, 2–5 mm. wide, membranous, ciliate, inner minutely pubescent, especially along the midrib, outer often glabrous, greenish to red-brown, persistent. Flowers dull yellow to orange-red, 4-merous; pedicel ± 1 mm. long, encircled by soft, white hairs. Calyx-tube 5–14 mm. long, glabrous below the articulation, minutely pubescent above; lobes oblong to ovate, obtuse to rounded, 2–3(–4) mm. long, (0·5–)1–2 mm. wide, minutely pubescent on the outside. Petals linear to club-shaped, 0·5–1 mm. long, rarely missing. Stamens subsessile; anthers 0·7–1 mm. long. Ovary glabrous; disc membranous, minute, cupular; style 3–6 mm. long. Seed 2–2·5 mm. long, 1–1·2 mm. wide.

TANZANIA. Ngara District: Busubi, Keza, 20 July 1960 (fl.), *Tanner* 5068!; Ufipa District: Rukwa, Sakalilo Escarpment, 1 Dec. 1954 (fl.), *Richards* 3502!; Chunya District: Lupa Forest Reserve, 22 Sept. 1962 (fl.), *Boaler* 687!
DISTR. T1, 4, 7; Mozambique, Malawi, Zambia, Rhodesia, South Africa (Transvaal, Natal), Swaziland
HAB. *Brachystegia* woodland and grassland subject to fire, sandy soil; 1300–2400 m.

SYN. *Arthrosolen pimeleoides* Meisner in DC., Prodr. 14: 560 (1857). Type: South Africa, Transvaal, Magaliesberg, *Burke* (G-DC, holo.!, NY-MEISN, iso.!)
Gnidia thomsonii H.H.W. Pearson in F.T.A. 6(1): 219 (1910) & in K.B. 1910: 337 (1910); T.T.C.L.: 608 (1949). Type: Tanzania, lower plateau N. of Lake Malawi, *Thomson* (K, holo.!)
Arthrosolen microcephalus (Meisner) Phillips in Journ. S. Afr. Bot. 10: 63 (1944), *non* S. Moore (1919)

5. **G. goetzeana** *Gilg* in E.J. 30: 363 (1901); H.H.W. Pearson in F.T.A. 6(1): 226 (1910); Domke in Bibl. Bot. 111, t. 3/27 (1934); Staner in B.J.B.B. 13: 350, fig. 12 (1935); T.T.C.L.: 608 (1949); F.F.N.R.: 272 (1962); A. Robyns in F.A.C. Thymelaeaceae: 43, fig. 3/D–F (1975). Type: Tanzania, Njombe District, Lumbira [Langenburg], by Rumbira R., *Goetze* 895 (B, holo. †, BR, lecto. !)

Erect, wiry shrub up to 1·5 m. high. Branches slender, glabrous, with brown or red bark. Leaves subsessile; leaf-blade narrowly elliptic or oblanceolate-linear, (7–)10–15(–24) mm. long, (1–)2–4(–6) mm. wide, glaucous green, glabrous. Inflorescence a terminal or axillary, bracteate head, 30–80-flowered. Bracts (4–)5–6(–7), ovate or ovate-lanceolate, acute to acuminate, 6–9 mm. long, 2·5–5 mm. wide, membranous, slightly sericeous inside, glabrous beneath, green, sometimes with reddish margins, deciduous in fruit. Flowers whitish or yellow, 4-merous; pedicel 1–2 mm. long, pubescent with a brush of hairs. Calyx-tube 5–7(–10) mm. long, densely but shortly silky-

hairy above the articulation, glabrous or shortly hairy below, especially in the
upper part; lobes ovate, rounded, 1–2 mm. long, 0·8–1·2 mm. wide, sparingly
pubescent outside. Petals 0·2–0·3 mm. long, filiform with granular tip or
glandular, sometimes rudimentary. Stamens subsessile; anthers 0·5–0·7 mm.
long. Ovary 1 mm. long, glabrous; style 1·5–3·5 mm. long; stigma papillate.
Seed ± 2 mm. long, 1 mm. wide. Fig. 6/5, p. 20.

TANZANIA. Mpanda District: Kungwe-Mahali Peninsula, between Pasagulu and
  Musenabantu, 10 Aug. 1959 (fl.), *Harley* 9278!; Iringa, just N. of township, 14 July
  1956 (fr.), *Milne-Redhead & Taylor* 11143!; Songea District: 3 km. NE. of Kigonsera,
  14 Apr. 1956 (fl.), *Milne-Redhead & Taylor* 9643!
DISTR. T4, 7, 8; Zaire, Burundi, Malawi, Zambia
HAB. *Brachystegia* woodland and wooded grassland; 800–1800 m.

SYN. *G. ramosa* H.H.W. Pearson in F.T.A. 6(1): 225 (1910) & K.B. 1910: 338 (1910).
  Types: Malawi, between Kondowe and Karonga, *Whyte* (K, syn.!, G, isosyn.!)
  & Chitipa [Fort Hill], *Whyte* (K, syn.!)
  Arthrosolen paludosa S. Moore in J.B. 57: 115 (1919). Type: Zaire, Shaba,
  Luente, *Kassner* 2485 (K, holo.!, BM, BR, P, iso.!)
  Gnidia marunguensis Staner in De Wild. & Staner, Contr. Fl. Katanga, Suppl. 4:
  71 (1932). Type: Zaire, Shaba, Kasiki, *de Witte* 404 (BR, holo.!)
  G. goetzeana Gilg var. *marunguensis* (Staner) Staner in B.J.B.B. 13: 352 (1935)

6. **G. stenophylla** *Gilg* in E.J. 19: 259 (1894) & in P.O.A. C: 283, t. 32/G–J
(1895); H.H.W. Pearson in F.T.A. 6(1): 226 (1910); T.T.C.L.: 608 (1949).
Type: Tanzania, Lushoto District, Usambara Mts., Kwa Mshusa, *Holst* 8963
(B, holo. †, K, lecto.!, G, P, S, iso.!)

Undershrub with numerous, slender shoots from a thick, woody rhizome, up
to 40–60 cm. high, unbranched or branched. Branches glabrous. Leaves
subsessile; leaf-blade needle-like, narrowly lanceolate, margins sometimes
slightly inrolled, 6–18 mm. long, 0·5–1·5 mm. wide, erect or ± spreading,
sometimes slightly curved, glabrous. Inflorescence terminal, involucrate,
10–30-flowered. Bracts (4–)5–8(–9), narrowly lanceolate to ovate, 3–8 mm.
long, 1–4 mm. wide, foliaceous or scarious, glabrous or slightly hairy in the
middle on the lower side, slightly hairy or puberulous on the upper, greenish
or brownish, margins often membranous, reddish and erose. Flowers yellow
or white, 4(–5)-merous; pedicel 1–1·5 mm. long, with stiff, erect hairs up to
3 mm. long. Calyx-tube 7–10(–12) mm. long, densely pubescent above the
articulation, pilose or villous below; lobes ovate to ovate-oblong, often somewhat emarginate, 1–3 mm. long, 0·5–2 mm. wide, puberulous outside. Petals
linear or spathulate, 0·5–1 mm. long, membranous or fleshy. Anthers subsessile, 0·5–1 mm. long. Ovary glabrous, 1–1·5 mm. long; style 2–7 mm. long;
stigma papillate. Seed 2–2·5 mm. long, ± 1 mm. wide. Fig. 6/6, p. 20.

UGANDA. W. Nile District: Koboko, Feb. 1934 (fl.), *Eggeling* 1426! & Apr. & Oct.
  1938 (fl.), *Hazel* 519! & 690! & Sept. 1940 (fl.), *Purseglove* 1037!
KENYA. Northern Frontier Province: Moyale, 26 Oct. 1952 (fl.), *Gillett* 14104!; Teita
  Hills, (fl.), *H.M. Gardner* 2955! & 13 Nov. 1963 (fl.), *Rauh* 12508!
TANZANIA. Bukoba District: on Msera–Namasina road, 16 June 1913 (fl.), *Braun* 5516!;
  Lushoto District: Lushoto–Mombo road, NE. of Vuga turnoff, 14 June 1953 (fl.),
  Drummond & Hemsley 2914!; Iringa District: 65 km. S. of Iringa on Mbeya road,
  14 Sept. 1958 (fl.), *Napper* 885!
DISTR. U1; K1, 7; T1, 3, 6–8; Cameroun, Zaire, Rwanda, Burundi, Ethiopia
HAB. Grassland subject to burning and wooded grassland, sometimes black soils;
  450–2100 m.

SYN. *G. urundiensis* Gilg in E.J. 51: 230 (1914); Staner in B.J.B.B. 13: 358, fig. 17
  (1935); A. Robyns in F.A.C. Thymelaeaceae: 48, fig. 5 (1975). Type: Burundi,
  H. *Meyer* 1108 (B, holo.†, BR, iso.!)
  G. cluytioides E.A. Bruce in K.B. 1933: 143, as " clutyoides ", 256 (1933). Type:
  Tanzania, Ngara/Biharamulo Districts, Karagwe, near Ruvuvu R., *Scott Elliot*
  8128 (K, holo.!)
  G. claessensii Staner in B.J.B.B. 13: 355, fig. 15 (1935); Aymonin in Fl. Cameroun
  5: 56, t. 7/3, 8/9–12 (1966) & Fl. Gabon 11, t. 15/3 (1966); A. Robyns in F.A.C.

Thymelaeaceae: 49, fig. 6 (1975). Type: Zaire, Orientale, Aru, *Claessens* 1555 (BR, holo.!)

NOTE. No characters hold to separate *Gnidia urundiensis*, *G. cluytioides* and *G. claessensii* from *G. stenophylla*. After having studied a rather large number of specimens I find these species should all be amalgamated. *G. stenophylloides* Gilg (endemic in Ethiopia) certainly also belongs to this difficult group.

7. **G. involucrata** *A. Rich.*, Tent. Fl. Abyss. 2: 234 (1850); Engl. in V.E. 1(1), fig. 132/b (1910); H.H.W. Pearson in F.T.A. 6(1): 221 (1910); Gastaldo in Webbia 24: 347, fig. 3, 6/B (1969). Types: Ethiopia, Tigre, Sana to Terrfera, *Schimper* 770 (P, lecto.!, BR, G, G-DC, K, P, S, isolecto.!) & Shire [Chiré], *Quartin Dillon & Petit* (P, syn.!, K, isosyn.!)

Perennial herb or undershrub, unbranched to much branched, up to 1 m. high or more, with a woody rhizome. Stems and branches glabrous, green to reddish, often brownish. Leaves sessile or shortly petioled; leaf-blade linear to elliptic, obtuse to acute, (7–)12–20(–25) mm. long, (1–)2–6(–8) mm. wide, glaucous, sometimes reddish, glabrous, young leaves rarely with scattered hairs. Inflorescence a terminal or axillary, (6–)15–25(–40)-flowered head. Bracts (4–)5–6(–8), elliptic to broadly oblong or subrotund, acute or rounded, 4–8(–10) mm. long, 2–7 mm. wide, leaf-like or scarious, glabrous, often reddish to brown, persistent. Flowers orange-yellow or pink to red, 4(–rarely 5)-merous; pedicel 0·5–1·5 mm. long, with a brush of silky hairs up to 3 mm. long. Calyx-tube (6–)8–12(–14) mm. long, glabrous to densely pubescent above the articulation, glabrous below; lobes oblong to ovate, (1–)2–3(–4) mm. long, 0·5–2 mm. wide, glabrous to hairy on the outside. Petals obovate to spathulate, 0·7–1·5 mm. long, 0·3–1·2 mm. wide, often ± emarginate, fleshy. Stamens subsessile; anthers 0·5–1 mm. long. Ovary glabrous, 1–1·5 mm. long; disc minute, cup-shaped, ± lobed; style 2–4 mm. long, excentric. Stigma papillate. Seed 2–3 mm. long, 1–1·5 mm. wide. Fig. 6/7, p. 20.

UGANDA. Acholi/Karamoja District: Lonyili Mts., Apr. 1960 (fl.), *J. Wilson* 996!
KENYA. W. Suk District: Kapenguria, 11 Jan. 1931 (fl.), *Lugard* K5!; Machakos/Masai District: Chyulu Hills, Apr. 1938 (fl.), *Bally* 11443!; Teita District: Bura Hills, *Gardner* in F.D. 2954!
TANZANIA. Ngara District: Busubi, Keza, 20 Nov. 1960 (fl.), *Tanner* 5607 A!; Kondoa District: Bereku, 17 Jan. 1928 (fl.), *B.D. Burtt* 1151!; Njombe District: between Lisitu and Lugalawa, 23 Sept. 1970 (fl.), *Thulin & Mhoro* 1152!
DISTR. U1; K2, 3, 4/6, 7; T1–8; N. Nigeria, N. Cameroun, Zaire, Sudan, Central African Empire, Ethiopia, Mozambique, Malawi, Zambia, Rhodesia, Angola
HAB. Open and wooded grassland, often in places subject to burning, also in deciduous woodland and bushland; 1000–2700 m.

SYN. *G. macrorrhiza* Gilg in E.J. 19: 260 (1894); H.H.W. Pearson in F.T.A. 6 (1): 218 (1910); Staner in B.J.B.B. 13: 347 (1935); F.P.S. 1: 148 (1950); F.W.T.A., ed. 2, 1: 174 (1954); F.F.N.R.: 272 (1962); Aymonin in Fl. Cameroun 5: 59, t. 7/1, 9/7–10 (1966); U.K.W.F.: 161 (1974); A. Robyns in F.A.C. Thymelaeaceae: 40 (1975). Type: Angola, Malange, *Mechow* 202 (B, holo. †, G, W, iso.!)
G. mittuorum Gilg in E.J. 19: 260 (1894); H.H.W. Pearson in F.T.A. 6(1): 218 (1910); F.W.T.A. 1: 151 (1927). Types: Sudan, Mittu, Deragoh, *Schweinfurth* 2850 (B, syn.†) & 2826 (B, syn.†, P, S, isosyn.!) & Reggo, *Schweinfurth* 2790 (B, syn.†, P, S, isosyn.!) & Dar Fertit, Dem Adlau, Dschih, *Schweinfurth*, ser. III, 114 (B, syn.†)
G. schweinfurthii Gilg in E.J. 19: 261 (1894); H.H.W. Pearson in F.T.A. 6(1): 219 (1910); F.P.S. 1: 148 (1950). Type: Sudan, Bahr el Ghazal, Sabbi, *Schweinfurth* 2851 (B, holo.†, K, iso.!)
G. buchananii Gilg in E.J. 19: 261 (1894); H.H.W. Pearson in F.T.A. 6(1): 219 (1910); Staner in B.J.B.B. 13: 348, fig. 11 (1935); T.T.C.L.: 607 (1949); Peterson in Bot. Notis. 111: 624 (1958) & in Bol. Soc. Brot., sér. 2, 33: 208 (1959); F.F.N.R.: 270 (1962); U.K.W.F.: 161 (1974); A. Robyns in F.A.C., Thymelaeaceae: 41, fig. 3/A–C (1975). Types: Malawi, *Buchanan* 125 (G, K, isosyn.!) & 179 (K, isosyn.!) & Blantyre, *Descamps* (BR, isosyn.!) & Chibisa to Tsinmuze, *Kirk* (B, syn.†, K, isosyn.)
G. leiantha Gilg in E.J. 19: 261 (1894); H.H.W. Pearson in F.T.A. 6(1): 217 (1910); T.T.C.L.: 608 (1949). Type: Tanzania, Dodoma District, Unyamwezi [Uniamwesi], Lake Chaya [Tschaia], *Stuhlmann* 433 (B, holo.†)

G. huillensis Gilg in E.J. 23: 206 (1896); H.H.W. Pearson in F.T.A. 6(1): 218
(1910). Type: Angola, Huila, *Antunes* 107 (B, holo.†)*
G. ringoetii De Wild. & Ledoux in De Wild., Contr. Fl. Katanga, Suppl. 2: 79
(1929). Types: Zaire, Shaba, Nieuwdorp, *Ringoet* 8 & without locality, *Verick*
(both BR, syn. !)

NOTE. This is a very variable species with series of intermediates between short
monocephalous specimens up to bushy, much branched specimens. The differences
are related to seasonal variation and severity of burning.

8. **G. usafuae** *Gilg* in E.J. 30: 363 (1901); H.H.W. Pearson in F.T.A. 6(1):
222 (1910); T.T.C.L.: 609 (1949). Type: Tanzania, Mbeya/Chunya Districts,
slopes of Poroto Mts., Usafwa, *Goetze* 1042 (B, holo. †, E, G, K, P, iso.!)

Much-branched shrub up to 1·5 m. high. Branches glabrous; bark brown
or chestnut-red. Leaves with petiole ± 1 mm. long; leaf-blade linear-
lanceolate, (8–)12–18(–26) mm. long, 1–8 mm. wide, glabrous. Inflorescence
capitate, terminal and axillary, 30–60-flowered. Bracts 4–6, broadly elliptic
to suborbicular, 6–9 mm. long, 5–8 mm. wide, scarious, glabrous, brown or
chestnut-coloured, persistent. Flowers yellow, 4-merous; pedicel 1·5–2 mm.
long, villous. Calyx-tube 6–10 mm. long, silvery pubescent above the articulation, glabrous below; lobes oblong, rounded, 1·5–2·5 mm. long, 1–1·2
mm. wide, pubescent on the outside. Petals linear or narrowly elliptic, 0·5–1·2
mm. long. Stamens with very short filaments, anthers 0·5–0·7 mm. long, the
upper whorl slightly exserted. Ovary glabrous; disc none; style 4–5 mm.
long. Seed 2·5–3 mm. long, 1 mm. wide.

TANZANIA. Mbeya Range, 16 Mar. 1960 (fl.), *Kerfoot* 1688!; Rungwe Mission, Jan.
1954 (fl.), *Semsei* 1541!; Njombe District: 13 km. WNW. of Njombe, Nyumbanyito,
11 July 1956 (fl.), *Milne-Redhead & Taylor* 11115!
DISTR. **T7**; Mozambique, Malawi
HAB. Upland grassland and forest margins; 1800–2300 m.

SYN.   *G. nutans* H.H.W. Pearson in F.T.A. 6(1): 221 (1910) & in K.B. 1910:337 (1910);
Peterson in Bol. Soc. Brot., sér. 2, 33: 213 (1959). Types: Mozambique,
mountains E. of Lake Malawi, *W.P. Johnson* (K, syn.!) & Malawi, between
Kondowe and Karonga, *Whyte* 330 (K, syn.!) & Nyika Plateau & S. Nyika
Mts., *Whyte* (both K, syn.!)

9. **G. chrysantha** (*Solms-Laub.*) *Gilg* in E.J. 19: 258 (1894); Staner in
B.J.B.B. 13: 363 (1935); F.P.S. 1: 148 (1950); Peterson in Bol. Soc. Brot., sér.
2, 33: 210 (1959); Aymonin in Fl. Cameroun 5: 68, t. 7/2, 11 (1966); Gastaldo
in Webbia 24: 351, fig. 4/D, 5/C–D (1969); U.K.W.F.: 159 (1974); A. Robyns
in F.A.C. Thymelaeaceae: 57 (1975). Type: Sudan, Blue Nile, Dali–Sennar,
Cienkowski (B, holo. †); Ethiopia, Kaffa, near Nadda, *Mooney* 6257 (K, neo.!)

Suffrutescent shrub, 20–50(–75) cm. high, with numerous simple or sparingly branched shoots from a woody rhizome. Leaves with petiole 0·5–1 mm.
long; leaf-blade linear to oblanceolate, (5–)8–18(–25) mm. long, 1–4(–6·5) mm.
wide, midrib prominent on the lower side, glabrous. Inflorescence a terminal,
involucrate head with 30–50 flowers; peduncle 3–30 mm. long. Bracts
(5–)6–8(–10), linear to ovate, outer narrower, 5–10(–17) mm. long, 1·5–4(–8)
mm. wide, foliaceous, glabrous, green, sometimes edged red. Flowers bright
yellow to red, 4-merous; pedicel 1–3 mm. long, puberulous with a ring of erect
hairs above the middle. Calyx-tube 9–16 mm. long, 4-angled, sericeous or
rarely glabrous above the articulation, glabrous below; lobes ovate-lanceolate
or obovate, 2·5–4 mm. long, 1·5–3 mm. wide, puberulous beneath. Petals
none. Stamens subsessile; anthers of the upper row ± 1 mm. long, enclosed

---

* A specimen in Coimbra (!) collected by *Antunes* but without No. 107 is named
G. huillensis by Gilg and may be an isotype.

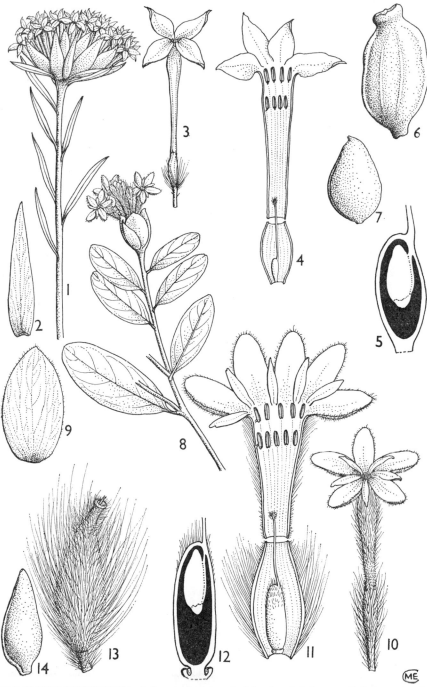

FIG. 7. *GNIDIA CHRYSANTHA*—**1**, flowering branch, × 1; **2**, bract, × 3; **3**, flower, × 3; **4**, flower, opened out, × 4; **5**, longitudinal section of ovary, × 24; **6**, fruit, × 12; **7**, seed, × 12. *G. EMINII*—**8**, flowering branch, × 1; **9**, bract, × 3; **10**, flower, × 3; **11**, flower, opened out, × 4; **12**, longitudinal section of ovary and disc, × 8; **13**, fruit, × 6; **14**, seed, × 6. 1–5, from *Milne-Redhead & Taylor* 9061; 6, 7, from *Lye* 6246; 8–12, from *Renvoize & Abdallah* 2316; 13, 14, from *Milne-Redhead & Taylor* 11172.

or slightly protruding, anthers of the lower row 0·6–0·8 mm. long. Ovary 1–1·5 mm. long, glabrous; disc none; style 3–5·5 mm. long, lateral; stigma papillate. Seed 1·5–2 mm. long, ± 1 mm. wide. Fig. 7/1–7.

UGANDA. Karamoja District: Mt. Napak, Apr. 1964 (fl.), *J. Wilson* 1680!; Masaka District: Katera, Oct. 1945 (fl.), *Eggeling* 5560! & Kyebe, Aug. 1945 (fl.), *Purseglove* 1777!
KENYA. Trans-Nzoia District: Kitale waterworks, Aug. 1931 (fl.), *Jex-Blake* in *Napier* 1402! & 6 km. on Kitale–Kapenguria road, 14 June 1958 (fl.), *Symes* 382!
TANZANIA. Bukoba District: Kakindu, Oct. 1931 (fl.), *Haarer* 2326!; Tabora District: Kakoma, 10 Aug. 1938 (fl.), *Glover* 180!; Songea District: by R. Ruvuma, near Kitai, 17 Apr. 1956 (fl.), *Milne-Redhead & Taylor* 9061!
DISTR.   U1, 4; K3; T1, 4, 6–8; Guinée, N. Nigeria, Cameroun, Zaire (Shaba), Sudan, Ethiopia, in the south to Rhodesia and Mozambique
HAB.   Seasonally wet grassland, less often in upland grassland subject to burning or in woodland; 400–2100 m.

SYN.   *Arthrosolen chrysanthus* Solms-Laub. in Schweinfurth, Beitr. Fl. Aethiop.: 165 (1867); H.H.W. Pearson in F.T.A. 6 (1): 234 (1910); T.T.C.L.: 606 (1949); F.W.T.A., ed. 2, 1: 176 (1954); F.F.N.R.: 270 (1962)
   A. glaucescens Oliv. in J.L.S. 15: 96 (1876). Type: Tanzania, Kigoma District, S. of Kawele, *Cameron* (K, holo.!)
   A. flavus Rendle in Trans. Linn. Soc., Bot., ser. 2, 4: 40 (1894). Types: Malawi, Mulanje [Mlanje] Mt., *Whyte* 99 (BM, syn.!) & Blantyre, *L. Scott* (K, syn.!)
   Gnidia stenosiphon Gilg in E.J. 19: 258 (1894). Type: Sudan, Dar-Fertit, Char Okilleah, W. of Pango, *Schweinfurth*, ser. III, 113 (B, holo.†)
   G. flava (Rendle) Gilg in E.J. 19: 258 (1894)
   G. ignea Gilg in E.J. 19: 258 (1894). Type: Tanzania, Tabora/Mpanda District, Boze at Ugalla R., *Böhm* 48a (B, holo.†)
   G. katangensis Gilg & Dewèvre in E.J. 19: 276 (1894); H.H.W. Pearson in F.T.A. 6(1): 226 (1910). Type: Zaire, Shaba, Katete, *Descamps* (BR, holo.!)
   G. glaucescens (Oliv.) Gilg in P.O.A. C: 283 (1895)
   Arthrosolen chrysanthus Solms-Laub. var. *ignea* (Gilg) H.H.W. Pearson in F.T.A. 6(1): 235 (1910); T.T.C.L.: 607 (1949)
   A. sphaerantha H.H.W. Pearson in F.T.A. 6(1): 235 (1910) & in K.B. 1910: 338 (1910); T.T.C.L.: 607 (1949). Type: Tanzania, Buha/Kigoma District, Malagarasi R., *Trotha* 56 (B, holo.†, K, fragm.!)
   Gnidia sphaerantha (H.H.W. Pearson) Gilg in V.E. 3(2): 633 (1921)
   G. chrysantha (Solms-Laub.) Gilg var. *ignea* (Gilg) Staner in B.J.B.B. 13: 365, fig. 20 (1935)

10.   **G. mollis** *C.H. Wright* in K.B. 1906: 23 (1906); H.H.W. Pearson in F.T.A. 6(1): 220 (1910); Staner in B.J.B.B. 13: 354, fig. 14 (1935); T.T.C.L.: 608 (1949); Peterson in Bol. Soc. Brot., sér. 2, 33: 212, t. I & II (1959); F.F.N.R.: 272 (1962); A. Robyns in F.A.C. Thymelaeaceae: 55 (1975). Types: Tanzania, lower and higher plateaux N. of Lake Malawi, *J. Thomson* & Mozambique, Uhangu to Lake Shire, *W.P. Johnson* 12 (all K, syn.!)

Shrublet with stout stem, virgately branched above the middle, up to 70 cm. (occasionally 2 m.) high. Branches densely pubescent, reddish brown, old stems and branches glabrous. Leaves with petioles 0·2–1 mm. long, sparsely hairy; leaf-blade lanceolate to oblong, acute to obtuse, 12–19 mm. long, 2–5 mm. wide, densely ciliate, shaggy pubescent beneath when young, later glabrescent. Inflorescence spherical, 20–25 mm. in diameter, compact, terminal or axillary, 50–100-flowered, composed of numerous small heads, each with 2–6(–12) flowers, surrounded by 4 membranous, pubescent, greenish to reddish bracteoles (abaxial ovate, 8–11 mm. long, 3–5 mm. wide; lateral lanceolate, 3–7 mm. long, 2–3 mm. wide; adaxial lanceolate, 3–5 mm. long, 2–3 mm. wide). Bracts 6–12, lanceolate to ovate, acute, 7–12 mm. long, 3·5–6 mm. wide, densely pubescent, greenish to slightly reddish, persistent. Flowers whitish or yellow, 4-merous; pedicel minute, with a brush of soft hairs. Calyx-tube 8–12 mm. long, glabrous below the articulation, densely shaggy-pubescent above; lobes oblong or ovate, obtuse, sometimes emarginate, ± 2 mm. long, 1–1·3 mm. wide, shaggy-pubescent outside. Petals linear, minute, 0·2–0·5 mm. long, or lacking. Stamens enclosed; anthers 0·5 mm. long. Ovary 1 mm. long, glabrous; disc very minute, lobed; style 4–6 mm. long; stigma papillate. Seed 2 mm. long, 1 mm. wide.

TANZANIA. Iringa District: 16 km. S. of Sao Hill, 16 Aug. 1949 (fl.), *Greenway* 8431! & Mufindi, May 1951 (fl.), *Eggeling* 6126!; Njombe District: by R. Hagafilo, 11 km. S. of Njombe, 8 July 1956 (fl.), *Milne-Redhead & Taylor* 10777!
DISTR. T7; Zaire (Shaba), Mozambique, Malawi, Zambia
HAB. Seasonally wet grassland and along water courses; 1400–2000 m.

SYN. *G. wittei* Staner in De Wild. & Staner, Contr. Fl. Katanga, Suppl. 4: 73 (1932). Type: Zaire, Shaba, Kasaki, *de Witte* 369 (BR, holo.!)

NOTE. The compound, compact inflorescence of this species is very characteristic. The same type of flower head is found in the closely related *Gnidia bambutana* Engl. (Cameroun and Nigeria) and *G. bakeri* Gilg (central Madagascar).

11. **G. kraussiana** *Meisner* in Hook., Lond. Journ. Bot. 2: 552, err. typ. "452" (1843) incl. vars.; Gilg in E.J. 30: 364 (1901); Staner in B.J.B.B. 13: 360, fig. 19 (1935); F.P.S. 1: 150, fig. 89 (1950); Peterson in Bot. Notis. 111: 627 (1958); Aymonin in Fl. Cameroun 5: 64, t. 7/4, 9/1–6 (1966) & in Bull. Soc. Bot. France 114: 221 (1967); U.K.W.F.: 159, fig. p. 160 (1974); A. Robyns in F.A.C. Thymelaeaceae: 50, fig. 7/A–C (1975). Type: South Africa, Natal, " ad latera mont. Tafelberge ", *F. Krauss* 455 (NY, holo.!, K, iso.!)

Perennial, unbranched or branched herb with herbaceous or lignous stems up to 60 cm. high from a thick woody base. Branches glabrous to densely pubescent. Leaves with petiole 1–2 mm. long; leaf-blade narrowly lanceolate or lanceolate, (13–)20–30 mm. long, 3–8(–10) mm. wide, to ovate, 30–45(–52) mm. long, (10–)15 20(–24) mm. wide, sometimes shortly mucronate, acute to obtuse, base tapering to rounded, larger leaves strongly nerved, glabrous to densely pubescent, often ciliate. Inflorescence capitate, terminal, 18 45(–72)-flowered; peduncle 1–12 cm. long, glabrous or pubescent. Bracts 5–10(–12), lanceolate to ovate, obtuse to acuminate, 10–18(–21) mm. long, 3–6 mm. wide, leaf-like, glabrous to pubescent, green, persistent. Flowers orange or yellow, (4–)5-merous; pedicel 1–3 mm., tomentose and with a fringe of soft hairs at top. Calyx-tube 7–12(–15) mm. long, densely pubescent above the articulation and with long sericeous hairs below; lobes lanceolate to ovate, rounded, usually ± emarginate, 2 4 mm. long, 1–2 mm. wide, densely pubescent on the outside, shining golden yellow inside. Petals very varied in shape, oblong to ovate, often ± emarginate, or lobed, 0·5–2 mm. long, 0·5–1 mm. wide, membranous, rarely glandular or lacking. Stamens subsessile, upper row usually exserted; anthers 0·5–1·5 mm. long. Ovary pilose to densely sericeous; disc minute, ring-shaped; style 4–11 mm. long; stigma papillate. Seed 2–2·5 mm. long, 1 mm. wide.

UGANDA. Acholi District: SE. Imatong Mts., Aringa R. Watershed, 7 Apr. 1945 (fl.), *Greenway & Hummel* 7312! & 2 km. N. of Lotuturu, 17 Feb. 1969 (fl.), *Lye* 2069; Ankole District: Karagwe, Aug. (fl.), *Scott Elliot* 8162!
KENYA. W. Suk District: Cherangani highway, Kapenguria, 29 Dec. 1960 (fl.), *Lind* 2845!; Trans-Nzoia District: Cherangani forest, Jan. 1935 (fl.), *Dale* in F.D. 3386!; Nandi District: Kaimosi, Apr. (fl.), *Archer* 241!
TANZANIA. Ngara District: near Keza, July 1953 (fl.), *Eggeling* 6661!; Singida District: Iramba Plateau, July 1958 (fl.), *Hammond* 55!; Masasi, Aug. 1965 (fl.), *Beecher* 93!
DISTR. U1, 2; K2, 3, 5; T1–5, 7, 8; Guinée, Togo, N. Nigeria eastwards to Sudan and southwards to Angola and South Africa (to Tembuland), Lesotho
HAB. Open grassland, commonly following recent burning, *Brachystegia*, *Uapaca* woodland, rocky banks; 1300–2700 m.

SYN. *Lasiosiphon kraussii* Meisner in DC., Prodr. 14: 596 (1857) incl. vars.; H.H.W. Pearson in F.T.A. 6(1): 231 (1910). Type as for *Gnidia kraussiana*
L. affinis Kotschy & Peyr., Fl. Tinn.: 39, t. 19/B (1867). Type: Sudan, W. of Dembo near Bongo, *Heuglin* 75 (W, holo.!)

Gnidia djurica Gilg in E.J. 19: 268 (1894) & in E. & P. Pf. III. 6a: 228 (1894), as " *dshurica* ". Types: Sudan, Jur, between Kutschuh Alis Seriba and Wau R., *Schweinfurth* 116 (B, syn.†, K, isosyn.!) & 117 (B, syn.†) & Dar Fertit, Golo Desert, *Schweinfurth* 115 (B, syn.†)

G. hoepfnerana Gilg in E.J. 19: 268 (1894). Type: South West Africa, Okahandja, *Höpfner* 40 (B, holo.†, G, iso.!)*

Lasiosiphon hoepfneranus (Gilg) H.H.W. Pearson in F.T.A. 6(1): 233 (1910)

Gnidia usinjensis Gilg in E.J. 19: 269 (1894); V.E. 3(2): 635 (1921), as " *usingensis* " pro syn. Type: Tanzania, Biharamulo/Mwanza District, Uzinza [Usinja], *Stuhlmann* 862 (B, holo.†)

Lasiosiphon kerstingii H.H.W. Pearson in F.T.A. 6(1): 233 (1910) & in K.B. 1910: 338 (1910). Type: Togo, Adjeidé, Kirikiri, *Kersting* 36 (B, holo.†, K, iso., fragm.!)

L. guineensis A. Chev., Expl. Bot. Afr. Occ. Fr. 1: 545 (1920), *nom. nud.*

Gnidia kerstingii (H.H.W. Pearson) Engl., V.E. 3(2): 636 (1921)

Lasiosiphon kraussianus (Meisner) Burtt Davy, Fl. Pl. & Ferns Transv. 1: 207 (1926); T.T.C.L.: 609 (1949); F.W.T.A., ed. 2, 1: 176, fig. 65 (1954); F.F.N.R.: 272 (1962); Aymonin in Bull. Soc. Bot. Fr. 114: 221 (1967)

L. kraussianus (Meisner) Burtt Davy var. *villosus* Burtt Davy, Fl. Pl. & Ferns Transv. 1: 46, pro specie, 207, pro var. (1926); T.T.C.L.: 609 (1949); I.T.U., ed. 2: 425 (1952). Type as for *Gnidia hoepfnerana*

NOTE. This species is the most variable of all species within *Gnidia*. It seems impossible to correlate the variations. *G. kraussiana* is also the species with the widest distribution within the African Thymelaeaceae.

Concerning the poisonous effects of this species see F.W.T.A. App.: 42 (1937) and Verdc. & Trump, Common Pois. Pl. E. Afr.: 36 (1969).

12. **G. eminii** *Engl. & Gilg* in E.J. 19: 265 (1894). Type: Tanzania, Kondoa District, W. of Irangi, *Stuhlmann* 1224 (B, holo. †)

Virgately branched shrub up to 3 m. high. Branches glabrous; bark dark grey or brown. Leaves falling off early; petiole ± 2 mm. long; leaf-blade oblong to oblong-lanceolate, 12–40 mm. long, 4–10 mm. wide, rigid, glabrous. Inflorescence terminal or axillary, 5–10-flowered head. Bracts 4–6, broadly ovate, rounded, 7–12 mm. long, 6–10 mm. wide, membranous, glabrous except for the finely ciliate margin, pale green, sometimes tinged brown or red-brown at the tip, persistent. Flowers orange to brick red, 5-merous; pedicel ± 2 mm. long, with long, stiff white hairs. Calyx-tube 13–15 mm. long, densely sericeous below the articulation with long erect hairs; lobes oblong, 2·5–4·5 mm. long, 1–1·5 mm. wide, sericeous on the lower side. Petals linear, thinly membranous, 2–3·5 mm. long. Stamens subsessile; anthers 1–1·2 mm. long. Ovary shortly stipitate, slightly hairy; disc small, cupular, 0·2–0·3 mm. high; style 3–4 mm. long. Seed 5 mm. long, 1·5–2 mm. wide. Fig. 7/8–14, p. 27.

TANZANIA. Shinyanga District: Mantini Hills, 13 May 1931 (fl.), *B.D. Burtt* 2455!; Singida District: Iramba Plateau, 6 km. on Kiomboi–Kasiriri road, 29 Apr. 1962 (fl.), *Polhill & Paulo* 2234!; Iringa District: 49 km. N. of Iringa, 16 July 1956 (fl., fr.), *Milne-Redhead & Taylor* 11172!

DISTR. T1, 2, 5–7; not known elsewhere

HAB. Deciduous bushland and thicket, often on rocky hills; 1000–1700 m.

SYN. *G. fischeri* Engl. & Gilg in E.J. 19: 266 (1894). Type: Tanzania, Kondoa District, Irangi, *Fischer* 542 (B, holo.†)

G. stuhlmannii Gilg in E.J. 19: 266 (1894) & in E. & P. Pf. III. 6a, fig. 78/F & G (1894); Engler in N.B.G.B., App. 11: 11, fig. F & G (1903) & in V.E. 1(2), fig. 410/F & G (1910). Types: Tanzania, Mpwapwa, *Stuhlmann* 248 (B, syn.†) & Kilosa, Kidete, *Stuhlmann* 185 (B, syn.†, K, isosyn.!)

Lasiosiphon eminii (Engl. & Gilg) H.H.W. Pearson in F.T.A. 6(1): 229 (1910); T.T.C.L.: 609 (1949)

L. fischeri (Engl. & Gilg) H.H.W. Pearson in F.T.A. 6(1): 229 (1910)

---

* According to Prodr. Fl. S. W. Afr. 86: 2 (1968) the type specimen of *G. hoepfnerana* is not collected at Okahandja (OK) but at a place with the same name in Sandfeld, Gobabis.

13. **G. latifolia** (*Oliv.*) *Gilg* in P.O.A. C : 283 (1895) ; Gastaldo in Webbia 24 : 365, fig. 8/4 (1969). Types : Tanzania, Kilimanjaro, *H.H.Johnston* (K, lecto. !), Kenya, Ribe to Galla country, *Wakefield* (K, syn. !)

Large, much-branched shrub up to 5 m. high. Young branches pubescent, later glabrous. Leaves with petioles 1–2 mm. long; leaf-blade oblong to oblanceolate, obtuse or rounded at the apex, 18–55 mm. long, 5–13 mm. wide, sparsely appressed pubescent beneath, with scattered hairs or usually glabrous above. Inflorescence a 6–12-flowered head ; leafless peduncle 20–30 mm. long, sparsely pubescent, glabrescent. Bracts 4–6, oblong or elliptic, 10–16 mm. long, 4–6 mm. wide, ciliate, finely silvery-pubescent on the outside, very sparsely pubescent or glabrous inside, caducous. Flowers yellow or orange, 5-merous ; pedicel ± 2 mm., pubescent. Calyx-tube 13–16 mm. long, densely hairy above the articulation, 3–4 mm. long white hairs below the articulation ; lobes oblong to elliptic, rounded at the apex, 3–5 mm. long, 2–3 mm. wide, pubescent beneath. Petals linear, 0·2–0·5 mm. long, glandular or usually lacking. Anthers 1·5 mm. long, upper whorl slightly exserted. Ovary shortly stipitate, the upper part hairy ; disc cupular, 0·3 mm. ; style 6–8 mm. long ; stigma papillate. Seed 4–4·5 mm. long, ± 2 mm. wide.

KENYA. Northern Frontier Province : Boni Forest, Mararani, 25 Dec. 1946 (fl.), *J. Adamson* 295 in *Bally* 5986 ! ; Machakos District : Emali, 29 Oct. 1959 (fl.), *Napper* 1329 ! ; Kwale District : between Samburu and Mackinnon Road, 30 Aug. 1953 (fl.), *Drummond & Hemsley* 4046 !
TANZANIA. Moshi District : Rau, Jan. 1936 (fl.), *Bancroft* 28 ! ; Pare District : SW. Pare, near Vudee, 30 Jan. 1930 (fl.), *Greenway* 2076 ! ; E. Usambara Mts., Maramba–Kijango, Apr. 1935 (fl.), *Greenway* 4040 !
DISTR. **K**1, 4, 5, 7 ; **T**2, 3 ; Somalia
HAB. Deciduous, coastal and upland evergreen bushland, wooded grassland ; 50–2000 m

SYN. *Arthrosolen latifolius* Oliv. in Trans. Linn. Soc., Bot. ser. 2, 2 : 348 (1887)
    Lasiosiphon hildebrandtii Engl., Hochgebirgsfl. Trop. Afr. : 310 (1892) *nom. nud.*, *non* Scott Elliot (1891)
    Gnidia vatkeana Engl. & Gilg in E.J. 19 : 267 (1894) ; Chiov., Fl. Somala 2 : 382, fig. 217 (1932). Types : Kenya, Kwale District, between Duruma and Teita, *Hildebrandt* 2369 (B, syn.†, K, P, isosyn. !) & Kitui District, Kitui, *Hildebrandt* 2838 (B, syn.†, G, K, P, W, isosyn. !) & Tanzania, Lushoto District, Mlalo, *Holst* 541 (B, syn.†) & Lutindi, *Holst* 3449 (B, syn.†, G, K, P, W, isosyn. !) & Lushoto/Tanga, Nyika steppe, *Holst* 2415 (B, syn.†, K, isosyn. !)
    Lasiosiphon vatkei H.H.W. Pearson in F.T.A. 6(1) : 228 (1910) ; [Engl. in E.J. 17 : 167 (1893), *nom. nud.* ;] T.S.K., ed. 2 : 18 (1936). Types as *Gnidia vatkeana*
    L. latifolius (Oliv.) Brenan in K.B. 4 : 93 (1949) ; T.T.C.L. : 610 (1949) ; K.T.S. : 556 (1961)

NOTE. Introduced into the Nairobi Arboretum, e.g. *G.R. Williams* 415, but apparently not commonly grown.

14. **G. lamprantha** *Gilg* in E.J. 19 : 264 (1894) ; P.O.A. C, t. 32/D–F (1895) ; F.P.S. : 150 (1950) ; Gastaldo in Webbia 24 : 362, fig. 8/B (1969). Types : Tanzania, Bukoba District, Karagwe, *Stuhlmann* 3204 (B, syn. †, K, lecto. !) & 1979 (B, syn. †, K, isosyn. !)

Much-branched shrub or small tree up to 3(–5) m. high. Branches densely tomentose. Leaves subsessile ; leaf-blade lanceolate to elliptic or oblanceolate, acute or shortly apiculate, 20–45 mm. long, 5–12 mm. wide, young leaves ciliate, pubescent (especially the lower part beneath), glabrescent. Inflorescence a dense, shortly peduncled, terminal head, 40–70-flowered. Bracts 6–10, ovate, 6–12 mm. long, 4–9 mm. wide, densely tomentose on both sides, persistent. Flowers yellow to orange-yellow, 5-merous ; pedicel short, with long, stiff hairs. Calyx-tube 8–12 mm. long, densely clothed with yellowish pubescence above the middle, dense tufts of silky hairs at the lower part ; lobes ovate or elliptic, obtuse, 3–4 mm. long, 1·5–3 mm. wide, pubescent beneath. Petals filiform, linear or linear-spathulate, 1–1·5 mm. long, mem-

branous. Stamens subsessile, the upper whorl exserted; anthers 1–1·2 mm.
long. Ovary pubescent at top, glabrous in the lower half; disc cupular; style
4–8 mm. long, stigma capitate. Seed 2·5–3 mm. long, 1–1·5 mm. wide.

UGANDA. Acholi District: Chua, Agoro, 15 Nov. 1945 (fl.), *A.S. Thomas* 4384!;
Ankole District: Ndeizha, 25 Apr. 1941 (fl.), *A.S. Thomas* 3832!; Mbale District:
Bugisu, Sipi, 30 Aug. 1932 (fl.), *A.S. Thomas* 411!
KENYA. Trans-Nzoia/N. Kavirondo District: Elgon, 26 Jan. 1931 (fl.), *Lugard* 516! &
13 Nov. 1957 (fl.), *Symes* 238!; Kericho District: Belgut Reserve, Cheptuiyet, 26
Aug. 1960 (fl.), *Kerfoot* 2186!
TANZANIA. Bukoba District: Karagwe, Feb. 1891 (fl.) & Apr. 1892 (fl.), *Stuhlmann*
1979 & 3204!
DISTR. U1–3; K3, 5; T1; Sudan, Ethiopia
HAB. Wooded grassland and bushland, often in rocky places; 1050–2100 m.

SYN. *Lasiosiphon lampranthus* (Gilg) H.H.W. Pearson in F.T.A. 6(1): 233 (1910);
T.S.K., ed. 2: 18 (1936); T.T.C.L.: 609 (1949); I.T.U., ed. 2: 425 (1952);
K.T.S.: 556 (1961)

15. **G. glauca** (*Fresen.*) *Gilg* in E.J. 19: 265 (1894); Staner in B.J.B.B. 13:
359, fig. 18 (1935); F.P.S. 1: 150 (1950); Aymonin in Fl. Cameroun 5: 69, t. 12
(1966); Gastaldo in Webbia 24: 358, fig. 7 (1969); A. Robyns in F.A.C.
Thymelaeaceae: 60, t. 7 (1975). Type: Ethiopia, N. of Gondar, *Rüppel* (FR,
holo.!)

Large, much-branched shrub up to 3·5 m. high or small tree up to 15(–24) m.
high. Branches densely leafy in the upper part, branchlets finely pubescent
when young, later glabrescent; bark grey, brown or blackish, rugose.
Leaves subsessile; leaf-blade lanceolate to oblanceolate, (20–)30–60(–80) mm.
long, 6–20 mm. wide, rigid, glaucous, finely pubescent near the base, glabrescent. Inflorescences dense terminal heads, 20–50-flowered; peduncle
pubescent, widening towards the top. Bracts 6–12, ovate, 10–15 mm. long,
6–10 mm. wide, slightly coriaceous, cream-coloured or salmon pink, softly
tomentellus on both sides, persistent. Flowers orange or golden yellow, fad-
ing to brown, 5(rarely 4)-merous; pedicel 1–2·5 mm. long, tomentose. Calyx-
tube 10–13 mm. long, the lower part with dense tufts of 2–4 mm. long silky
hairs, upper part softly tomentose, no articulation; lobes ovate, 2·5–4 mm.
long, 1·5–2 mm. wide, tomentose outside. Petals spathulate, 1–2 mm. long,
entire, emarginate or lobed, membranous or fleshy. Stamens in upper row
slightly exserted; anthers 1–2 mm. long. Ovary pubescent especially at apex;
style 6–8 mm. long; stigma globose. Seed 3 mm. long, 1·5–2 mm. wide.

UGANDA. Acholi District: Imatong Mts., Agoro, *Eggeling* 1192!; Mbale District:
Elgon, Benet, Jan. 1936 (fl.), *Eggeling* 2464!
KENYA. Northern Frontier Province: Mt. Nyiru, 14 Feb. 1947 (fl.), *J. Adamson*
386 in *Bally* 6156!; N. Nyeri District: Mt. Kenya, Sirimon Track, 22 Sept. 1963 (fl.),
Verdcourt 3774!; Masai District: Olokurto, 13 May 1961 (fl.), *Glover, Gwynne &
Samuel* 937!
TANZANIA. Arusha District: Ngurdoto National Park, Tulusia Hill, 7 Nov. 1965 (fl.),
Greenway & Kanuri 12308!; Iringa District: Mufindi, Nyamalalu, May 1951 (fl.),
Eggeling 6080!; Songea District: Matengo Hills, Liwiri-Kiteza Forest Reserve,
4 Oct. 1956 (fl.), *Semsei* 2503!
DISTR. U1, 3; K1–6; T2, 4–8; Nigeria, Cameroun, Zaire, Sudan, Ethiopia, Malawi,
Zambia
HAB. Upland forest margins and associated bushland or wooded grassland; 1500–
3300 m.

SYN. *Lasiosiphon glaucus* Fresen. in Flora 21: 603 (1838); H.H.W. Pearson in F.T.A.
6(1): 230 (1910); T.S.K., ed. 2: 18 (1936); T.T.C.L.: 609 (1949); I.T.U., ed.
2: 424, t. 20 (1952); F.W.T.A., ed. 2, 1: 176 (1954); K.T.S.: 556, t. 31 (1961);
Archangelsky in Kuprianova, Pollen Morphology: 188, t. 15/22 (1971)
Gnidia volkensii Gilg in P.O.A. C: 283 (1895). Type: Tanzania, Kilimanjaro,
Nokolu, *Volkens* 2012 (B, holo.†, BM, E, G, K, LE, iso.!)

7. STRUTHIOLA

L., Syst. Nat., ed. 12, 2: 127 (1767) & Mant. Pl.: 4 (1767), *nom. conserv.*;
Domke in Bibl. Bot. 27(111): 135 (1934); G.F.P. 2: 259 (1967)
Belvala Adans., Fam. Pl. 2: 285 (1973), *nom. rejic.*

Erect shrubs or undershrubs of ericoid habit. Branches 4-angled or terete,
with usually prominent leaf scars. Leaves opposite, whorled in threes, or
alternate, sessile; blade simple, entire, linear to suborbicular, margins often
ciliated, coriaceous or subcoriaceous, parallel-veined. Inflorescence a spike.
Flowers 4-merous, sessile, usually in the axils of the upper leaves, solitary,
rarely in pairs, fragrant at dusk. Bracteoles 2, conduplicate, usually ciliate.
Calyx-tube cylindric, slightly widened at top, circumscissile above or below
the top of the ovary, glabrous or hairy; lobes 4, much shorter than the tube,
imbricate, the 2 outer slightly larger than the inner. Petals 4, 8 (in Flora
area) or 12, erect, fleshy, exserted, each surrounded by stiff hairs arising from
the base. Stamens 4, in one whorl, inserted in the throat of the calyx-tube,
subsessile, alternating with the calyx-lobes; anthers linear to oblong, basi-
fixed, included or slightly exserted; pollen sphaeroid, polyforate. Ovary
sessile or subsessile, 1-chambered, glabrous; disc surrounding the base of the
ovary, lacking or inconspicuous, cup-shaped; style filiform, inserted laterally,
half as long to almost as long as the calyx-tube; stigma minute, ± capitate,
papillose. Fruit dry, indehiscent, included in the persistent base of the calyx-
tube. Seed with shiny black, crustaceous testa, the micropyle forming a
curved beak; endosperm scanty; embryo straight; cotyledons fleshy.

About 30 species, mainly South African, two in tropical Africa.

S. thomsonii *Oliv.* in J.L.S. 21: 404 (1885) & in Hook., Ic. Pl. 15: 73, t. 1493
(1885); H.H.W. Pearson in F.T.A. 6(1): 215 (1910); Engl. in V.E. 3(2): 636
(1921); Fries in N.B.G.B. 8: 421 (1923); T.S.K., ed. 2: 17 (1936); T.T.C.L.:
611 (1949); Peterson in Bot. Notis. 111: 419 (1958); K.T.S.: 557 (1961);
Gastaldo in Webbia 24: 386, fig. 14/B & 15 (1969); U.K.W.F.: 159 (1974);
A. Robyns in F.A.C. Thymelaeaceae: 64, t. 8 (1975). Type: Kenya, Laikipia,
Thomson (K, holo.!)

Shrub or undershrub, sparsely to much branched, up to 2(3) m. high.
Branches 4-angled to terete, light to dark brown, densely leafy in the upper
part, pubescent, later glabrescent, gradually with prominent leaf scars.
Leaves whorled or opposite, at first imbricate, later ± spreading; blade
linear-lanceolate to ovate, (4–)6–15(–20) mm. long, 1·5–3(–6) mm. wide, acute
to subacute, margins sometimes inrolled, coriaceous or subcoriaceous, glabrous, ciliate when young. Flowers white, yellow or red. Bracteoles linear to
ovate-lanceolate, 2–5 mm. long, 0·5–1 mm. wide, ciliate. Calyx-tube slender,
glabrous, (5–)6–8(–14) mm. long; lobes lanceolate to ovate-lanceolate, 1·5–4
mm. long, 1–1·5 mm. wide at the base, acute to obtuse, the 2 exterior with a
small tuft of hairs at the tip. Petals 8, terete, 0·5–1 mm. long, fleshy. Anthers 0·5–1·2 mm. long, acute. Ovary oblong, 1–1·5 mm. long, glabrous; disc
minute; style up to 6 mm. long. Seed ± 2 mm. long, ± 1 mm. wide. Fig. 8.

KENYA. Northern Frontier Province: Mt. Nyiru, 31 Dec. 1955 (fl.), *J. Adamson* 545!;
    Naivasha District: Mt. Longonot, 8 Nov. 1959 (fl.), *Polhill* 58!; Masai District: Ol
    Doinyo Orok, 8 Dec. 1944 (fl.), *Bally* 4174!
TANZANIA. Arusha District: Ngurdoto National Park, Tulusia Hill, 7 Nov. 1965
    (fl.), *Greenway & Kanuri* 12312!; W. Usambara Mts., 3 km. NW. of Mlalo, Zevigambo,
    16 June 1953 (fl.), *Drummond & Hemsley* 2955!; Uluguru Mts., Lukwangule Plateau,
    31 Jan. 1935 (fl.), *E.M. Bruce* 690!
DISTR. K1–6; T2–4, 6; Zaire (Kivu), Rwanda, Burundi, S. Ethiopia
HAB. Upland grassland and moorland, often in the ericaceous zone; 1800–4000 m.

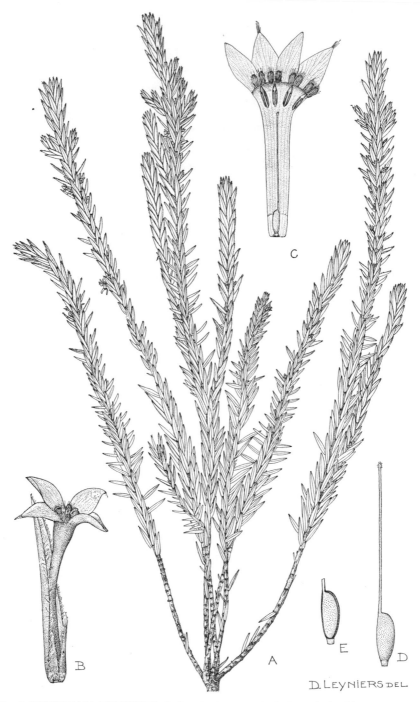

D. LEYNIERS DEL

FIG. 8. *STRUTHIOLA THOMSONII*—**A**, flowering branch, × ½; **B**, flower, leaf and bracteoles, × 5; **C**, flower, opened out, × 5; **D**, pistil, × 8; **E**, longitudinal section of ovary, × 8. A, from *Lewalle* 6001; B–E, from *Bamps* 3001. Reproduced by permission of the Director, Jardin Botanique National de Belgique, from " Flore d'Afrique Centrale ".

SYN. *S. usambarensis* Engl. in E.J. 19: 269 (1894) & in V.E. 1(1): 333, fig. 297/D–F
      (1910); H.H.W. Pearson in F.T.A. 6(1): 214 (1910); T.T.C.L.: 612 (1949).
      Type: Tanzania, Usambara Mts., Mlalo, Ngambo, *Holst* 41 (B, holo.†)
      S. ericina Gilg in E.J. 19: 270 (1894); Engler in V.E. 1(1): 333, fig. 297/A–C
      (1910); H.H.W. Pearson in F.T.A. 6(1): 214 (1910); T.T.C.L.: 611 (1949).
      Type: Tanzania, Usambara Mts., Mtai, Tewe Stream, *Holst* 2476 in part (B,
      holo.†, BR, COI, HBG, K, LE, M, P, W, Z, iso.!)
      S. stuhlmannii Gilg in P.O.A. C: 283 (1895); H.H.W. Pearson in F.T.A. 6(1):
      214 (1910); T.T.C.L.: 611 (1949). Type: Tanzania, Uluguru Mts., Lukwangule,
      Stuhlmann 9219 (B, holo.†)
      S. kilimandscharica Gilg in P.O.A. C: 284 (1895); H.H.W. Pearson in F.T.A.
      6(1): 215 (1910); T.T.C.L.: 611 (1949). Type: Tanzania, Kilimanjaro, N. side
      of Mawenzi Peak, R. Usseri, *Volkens* 2008 (B, holo.†, BM, G, K, iso.!)
      S. amabilis Gilg in P.O.A. C: 284 (1895); H.H.W. Pearson in F.T.A. 6(1): 215
      (1910); T.T.C.L.: 611 (1949). Type: Tanzania, Uluguru Mts., Lukwangule,
      Stuhlmann 9157 (B, holo.†)
      S. gilgiana H.H.W. Pearson in F.T.A. 6(1): 214 (1910) & in K.B. 1910: 336
      (1910); T.T.C.L.: 611 (1949). Type: Tanzania, Usambara Mts., slopes of
      Mtai, *Holst* 2476 in part (K, holo.!)
      S. albersii H.H.W. Pearson in F.T.A. 6(1): 215 (1910) & in K.B. 1910: 336
      (1910); T.T.C.L.: 611 (1949). Type: Tanzania, W. Usambara Mts., Kwai,
      Albers 191 (K, holo.!)
      S. volkensii H. Winkler in F.R. 9:524 (1911); T.T.C.L.: 612 (1949). Type:
      Tanzania, Kilimanjaro, above Moshi, *Winkler* 3989 (WRSL, holo.†?, BR, iso.!)

INDEX TO THYMELAEACEAE

GEOGRAPHICAL DIVISIONS OF THE FLORA

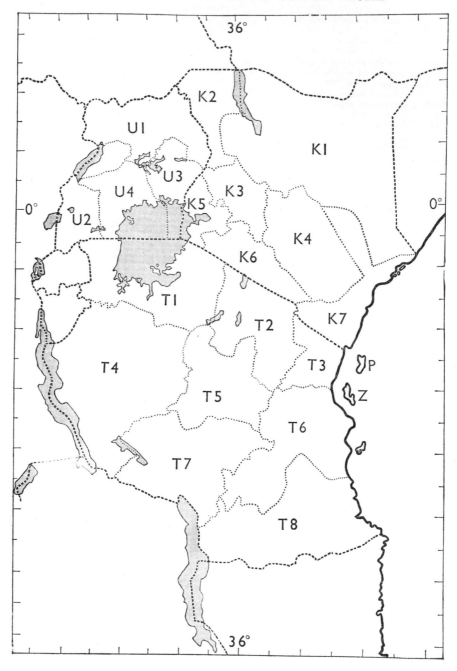